C0-AYQ-124

The Unfulfilled Promise of Synthetic Fuels

Recent Titles in
Contributions in Political Science
Series Editor: Bernard K. Johnpoll

Israel-Palestine: A Guerrilla Conflict in International Politics
Eliezer Ben-Rafael

Federal Lands Policy
Phillip O. Foss, editor

Begin's Foreign Policy, 1977–1983: Israel's Move to the Right
Ilan Peleg

Comparing Presidential Behavior: Carter, Reagan, and the Macho Presidential Style
John Orman

Toward a Theory of Eurocommunism: The Relationship of Eurocommunism to Eurosocialism
Armen Antonian

Power and Pluralism in American Cities: Researching the Urban Laboratory
Robert J. Waste

Class in China: Stratification in a Classless Society
Larry M. Wortzel

The White House Staff and the National Security Assistant: Friendship and Friction at the Water's Edge
Joseph G. Bock

Why Americans Don't Vote: Turnout Decline in the United States, 1960–1984
Ruy Antonio Teixeira

The Trampled Grass: Tributary States and Self-reliance in the Indian Ocean Zone of Peace
George W. Shepherd, Jr.

Neighborhood Justice in Capitalist Society: The Expansion of the Informal State
Richard Hofrichter

Presidential Power and Management Techniques: The Carter and Reagan Administrations in Historical Perspective
James G. Benze, Jr.

Tickamyer

THE UNFULFILLED PROMISE OF SYNTHETIC FUELS

Technological Failure, Policy Immobilism, or Commercial Illusion

Edited by ERNEST J. YANARELLA and WILLIAM C. GREEN

Contributions in Political Science, Number 179

GREENWOOD PRESS
New York • Westport, Connecticut • London

Library of Congress Cataloging-in-Publication Data

The Unfulfilled promise of synthetic fuels.

(Contributions in political science,
ISSN 0147–1066 ; no. 179)
Bibliography: p.
Includes index.
1. Synthetic fuels industry—United States.
I. Yanarella, Ernest J. II. Green, William C., 1941–
III. Series.
HD9502.5.S963U5475 1987 338.4′766266′0973 87–247
ISBN 0–313–25666–7 (lib. bdg. : alk. paper)

British Library Cataloguing in Publication Data is available.

Library of Congress Catalog Card Number: 87–247
ISBN: 0–313–25666–7
ISSN: 0147–1066

First published in 1987

Greenwood Press, Inc.
88 Post Road West, Westport, Connecticut 06881

Printed in the United States of America

The paper used in this book complies with the
Permanent Paper Standard issued by the National
Information Standards Organization (Z39.48–1984).

10 9 8 7 6 5 4 3 2 1

To

Ernest D. and Margaret M. Yanarella

and

Essington F. and Elizabeth C. Green

Contents

Illustrations

TABLES

FIGURES

Abbreviations

AEC	Atomic Energy Commission
AGA	American Gas Association
COED	Char Oil Energy Development
DEIS	Draft Environmental Impact Statement
DOE	Department of Energy
DPA	Defense Production Act
EC	European Communities
EIS	Environmental Impact Statement
ERDA	Energy Research and Development Administration
GOCO	Government-Owned, Corporation-Operated
GOGO	Government-Owned, Government-Operated
IEA	International Energy Agency
IMMR	Institute for Mining and Minerals Research
KCER	Kentucky Center for Energy Research
NATO	North Atlantic Treaty Organization
NCB	National Coal Board (United Kingdom)
NEPA	National Environmental Policy Act
NPC	National Petroleum Council
OCR	Office of Coal Research
OECD	Organization for Economic Cooperation and Development
OMB	Office of Management and Budget
OPEC	Organization of Petroleum Exporting Countries

R&D	Research and Development
RD&D	Research Development and Demonstration
RFP	Request for Proposals
SASOL	South African Coal, Oil, and Gas Corporation
SFC	Synthetic Fuels Corporation
SFT	Synthetic Fuels Technology
SRC	Solvent Refined Coal
TOSCO	The Oil Shale Corporation

Preface

ERNEST J. YANARELLA AND WILLIAM C. GREEN

The dismantlement of the Synthetic Fuels Corporation signals the failure of America's most recent effort to extract synthetic fuel from coal for commercial use. This latest chapter in the history of synfuel development in the United States has followed a now-familiar pattern in this nation's oil policies: a domestic shortfall or international disruption in oil supplies provokes alarmist calls to intensify energy research and development to achieve energy independence. A well-orchestrated coalition of corporate, political, and military actors attempts to mobilize public officials and the public behind a technological response to this energy crisis (really a liquid fuels crisis). American scientists and engineers formulate bold designs and make grandiose promises that arouse high public expectations. Politicians and policy makers dedicate themselves to the objective of energy autarky through this technological fix strategy while releasing massive subsidies from the federal treasury to support the latest technological alternative. Long-range programs are put in place and small-scale demonstration plants are built. Then, suddenly, the crisis subsides because of new or unexpected discoveries or changes in the international energy system. The commercial promise of the highly touted alternative technology fades; the public infrastructure is torn down; and pilot plants are mothballed.

This mobilization model is based on the history of American technological development in the twentieth century, but most or all of its elements have been featured in each of the preceding periods of synthetic fuels development in the twenties, the fifties, and now in the late seventies and early eighties. In the latest mobilization campaign, this model or strategy of policy innovation has obviously failed to generate the necessary political support and the requisite technological infrastructure to build the kind of energy future envisaged by the Carter administration's National Energy Plan II. With the dismemberment of the Synthetic Fuels Corporation and the shelving of scores of synfuel plant proposals by the federal and state government agencies and their industrial coventurers, the time is at hand to undertake a deep and searching inquiry into the reasons why the synthetic fuel bubble has burst.

This volume is designed to apply the critical tools of political, economic, and scientific analysis to this most recent effort to develop a commercially viable synthetic fuels industry. It grows out of our conviction that a substantial work on the subject will serve as a knowledge base for evaluating and drawing lessons from the failure of this mobilization campaign and for assessing the potential of synfuel technology. We make no attempt to reach a consensus on the ultimate value or place of synfuels in America's or the world's energy future. Policy pluralism rather than policy consensus informs the foundations of these scholarly essays. Some of the social scientists, policy analysts, and energy planners who have contributed to this volume view synthetic fuels as a legitimate and important object of political or social inquiry, a fact which reveals certain enduring patterns of policy stalemate, administrative reorganization, and legislative politics. Synfuels, for others, hold out the hope of carving a place, albeit a more modest one than previously touted, in the contours of the energy landscape of the twenty-first century. For still others, they are little more than commercial illusions that periodically dip into the public treasury and constantly pose environmental hazards and community disruptions in their technological advancement and commercial development. Overall, the contributors find that the politics and economics of synfuel development here and overseas are a fascinating, complex, and highly instructive phenomenon.

The two essays that comprise Part I of this volume focus on key historical and theoretical perspectives on synthetic fuels development in the United States. In "Business, Government, and Markets: Synthetic Fuels Policy in America," business historian Richard H. K. Vietor explores the series of federal programs formulated between 1944 and 1986 to stimulate commercial development of synthetic fuels. Analyzing these programs within a political economy framework, Vietor explains their failure by looking at certain changes in energy markets during the course of these developments, by examining the dynamic relationship between markets and politics, and by probing various institutional channels for business-government relations.

Vietor notes that the liquid fuels market was shaken by four pronounced shifts in the balance between supply and demand during this period. Using documentary evidence drawn from government files recently made available to scholars and the public, he turns to public-interest and private-interest theories of policy formulation to explain why these disruptions triggered powerful political pressures sometimes in favor, sometimes against government involvement in synthetic fuels development. Vietor concludes that during periods of real or perceived liquid fuels shortages when bureaucratic and congressional actors dominate efforts to devise a government program to alleviate the shortfall, the policy process is better characterized by the public-interest/collective-action theories identified by Mancur Olson and William Baumol. During periods when energy supplies are large and energy prices fall, Vietor finds that the private-interest theories of Gerald Nash, Gabriel Kolko, and K. Austin Kerr offer stronger explanations of the policy process and the self-interested behavior of firms and industry groups. Rich in historical detail and replete with theoretical insights, Vietor's analysis of the American politics of synthetic fuels nondevelopment provides the reader with a sure guide through the historical maze and political thicket of synthetic fuels politics.

Michael Crow is a political scientist and close student of the organizational politics of science and technology who grapples with the issue of "why the [synfuels] technology itself has failed to develop much beyond the first generation efforts of the 1920s and why the technology has continually been plagued by failure and cost overruns." In "Synthetic Fuel Technology Nondevelopment and the Hiatus Effect," Crow explores these shortcomings in terms of an assortment of explanations found in the policy literature: (1) fluctuations in the petroleum market—particularly dramatic swings in supply and price—have sabotaged efforts to stabilize the financial risks of synthetic fuels development and to uncouple the costs of synthetic fuels production from the price of oil; (2) the continuing conflict between the collective good and private interest has prevented cooperation between government agencies and industrial enterprises to develop unified and continuous technology development programs; (3) pork-barrel and interfuel politics, multiple policy goal setting, and political risk-reducing strategies have conspired to undercut the potential for synthetic fuel development; and (4) the "crisis" character of government involvement in synfuels has spawned a mobilization strategy that replaces an orderly, sequential development program with one involving simultaneous research and development on diverse facets of the technology, thus maximizing technological risks and cost uncertainties.

Crow reviews each of these plausible explanations and notes the partial nature of their explanatory power. On the basis of his close scrutiny of synfuels and other macro-engineering projects, he explains the failure of synfuels development in the United States in terms of the hiatus effect. According to this hypothesis, the inability of synfuels technology to get much beyond the first generation and its production inefficiencies and cost overruns can be understood in terms of the tremendous losses of information, knowledge, and momentum that result from the hiatus periods between accelerated synfuels research and development programs. At the onset of each new mobilization campaign, so much of the organizational and personnel infrastructures must be reassembled that research and development must practically return to an earlier stage of technological development. The hiatus periods in synfuels development, Crow argues, have been especially devastating because this loss in continuity and momentum cannot easily or obviously be compensated for either by a mobilized policy environment or massive infusion of public capital. Moreover, this lurching character of synthetic fuels development means that coal technologists will not have the 20- to 25-year lead time necessary to phase in advanced clean coal technology. Existing inefficient synfuels technologies will provide the only means for meeting a future need. Therefore, Crow advocates an alternative strategy of initiating a realistic, long-term research and development program two or three decades before synthetic fuels must be incorporated into America's energy mix.

The six essays that comprise Part II examine state and national political perspectives on American synthetic fuels development during the recent mobilization effort. The first is Patrick Hamlett's "Technological Policy Making in Congress: The Creation of the U. S. Synthetic Fuels Corporation." In this study of the Energy Security Act of 1980, the statute which created the Synthetic Fuels Corporation (SFC), Hamlett describes how congressional supporters of this technology development program were able to exploit certain procedural devices to advance the fortunes of the energy bill and how H.R. 3930, while by no means assured of passage, succeeded in mustering wider circles of support among crucial House Democratic leaders until its opponents conceded defeat. Hamlett then examines the synergistic relationship that developed between the actions of the Senate supporters of a synthetic fuels bill and the campaign of President Carter to refashion his national energy program around the synfuels option. He carefully reconstructs the complicated legislative history of SFC as well as the surprising growth in the magnitude of the funding levels for synthetic fuels development. The involvement of powerful committees and influential senators and congressmen, coupled with the president's determination to overcome the nation's malaise through a sizable program, is also detailed with a sense of the complex process of policy innovation and legislative logrolling that shaped the passage of the Energy Security Act.

Hamlett finds the same parliamentary tactics, processes of coalition building, and negotiation evident in synfuels legislation as in legislation that focuses on social, fiscal, and other nonscientific and nontechnological issues. At the same time, he also finds one feature of the synfuels legislation that characterizes technological policy making: the American penchant to support high-technology efforts to solve national crises. Congress overlooked the long-term political, socioeconomic, and environmental problems associated with a crash synfuels development program in its stampede to foster rapid commercial development of synthetic fuels. It set aside numerous technical assessments by congressional research agencies, all skeptical of the bill's projected production goals and all alarmed by the likely environmental impacts, and, abetted by White House pressures, increased synfuels funding and enlarged the scope of the program to support huge federal outlays for research, development, and demonstration projects, in spite of pervasive legislative resistance to heavy federal involvement in the development of this technology. Thus, Hamlett concludes that, despite the tendency of legislative policy making to be slow, cumbersome, and seemingly byzantine, the case of the Energy Security Act and the creation of the Synthetic Fuels Corporation suggests that congressional representatives, no less than the American public, are prone to favor crash-program/mobilization strategies in perceived emergency situations as instruments for overcoming policy stalemate in the middle levels of power.

In "The Synthetic Fuels Corporation as an Organizational Failure in Policy Mobilization," Sabrina Willis contributes to our understanding of the failure of the federal agency established to accomplish the objectives of the most recent synfuels policy mobilization. Willis argues that the initial responsibility for the SFC's short life was due initially to the political and economic climate of opinion that pervaded the legislative process. The Energy Security Act showed little understanding of the steps necessary to promote the successful development of this high-technology energy alternative and established unrealistic goals for the SFC. She concentrates her attention on three other factors that account for the corporation's demise in late 1985 and early 1986: the president, the prevailing economic environment, and the corporation itself.

President Reagan had a long-standing hostility towards the Synthetic Fuels Corporation. Sharing the same pseudo–free-market philosophy as his OMB director David Stockman, an early congressional opponent of the SFC, the president did little to garner political support for the SFC or to support its legislative purpose or policy vision. The Synthetic Fuels Corporation also suffered from international economic developments. The sense of a crisis atmosphere that had initially triggered interest in the synfuels option quickly dissipated along with its political support when the powerful OPEC cartel dramatically weakened and global oil prices began to spiral downward from early 1985 onward. With the end of this mobilized environment, as Vietor observes, went the possibility of industry cooperation in synfuels development. For Willis, however, the SFC's personnel and program difficulties made the agency its own worst enemy. Unprofessional, unethical, or questionable conduct led to the resignation of two chief administrators. Moreover, the agency's strategy, which called for massive public investment in synfuels technologies, was based on the mistaken belief that the South African program proved the maturity of those technologies.

Could the SFC have been more successful in fulfilling its mandate by taking a different course? Willis believes that the agency was faced with an impossible political situation, but she also suggests that it could have done more by asserting its administrative autonomy and funding small-scale projects that would have expanded the scope of knowledge about the costs, risks, and benefits of generating an American synfuels industry. Given the low public esteem of synthetic fuels as a viable alternative to energy technology, she concludes that it is problematic whether synfuels will surmount the SFC debacle.

In "A State Government's Experience with the Synthetic Fuels Movement," John Mitchell, a professional engineer and energy planner with the Kentucky Energy Cabinet, offers an analysis of the Commonwealth of Kentucky's strategy for promoting synfuels technology in the seventies and early eighties. After an overview of the most recent national synthetic fuels development campaign, Mitchell argues that even without the Arab oil embargo of 1973–1974, the time seemed ripe for an accelerated program of exploiting America's most abundant energy resource through technological means: keen interest in coal technology was already at a high level, coal conversion technology appeared to be poised for large-scale development, and energy R&D functions had just been centralized in the U.S. Energy Research and Development Administration (ERDA). Kentucky state government had also undertaken an aggressive program to exploit its enormous coal reserves and to support its coal industry. After establishing a research and development infrastructure in the Kentucky Center for Energy Research and its high-technology laboratory in 1972, the General Assembly created the Energy Development and Trust Fund in 1974 to facilitate state participation in large-scale pilot and demonstration projects. Although this two-pronged program was unique among the states, Mitchell argues that Kentucky's aggressive pursuit of synfuels seemed warranted. Given the commonwealth's public commitment to coal use and coal conversion technology, it appeared that Kentucky coal would play a dominant role in national energy plans and policies through the end of the century and beyond.

Mitchell's review of his state's fourteen-year program in energy research and development leads him to conclude that Kentucky's balance sheet on energy is extremely mixed. Over $27 million has already been spent on seventeen large-scale synfuels projects. Cost-benefit calculations suggest to him that, had all these plants become operational, the economic benefits in jobs, coal use, and energy production would have been impressive. On the other hand, the failure of virtually all of these plants, including the only completed and operational project, the H-Coal Pilot Plant project at Catlettsburg, results in more sober conclusions. Some new knowledge of the design, operation, and use of Kentucky coal was gained, but the myriad benefits expected from the state's research and development program were not realized.

In conclusion, Mitchell asks: What was gained from the $572 million invested by all parties in the commonwealth's synfuels development program? His answer: Perhaps as much as $218 million was spent in Kentucky on labor, goods, and services; an energy research laboratory that has become a leader in advanced coal research was established; and several instrumentation spin-offs applicable to public health and socioeconomic impact analysis were developed. Mitchell also suggests that important advances in coal conversion technologies were accomplished during this last wave of synfuels mobilization. Still, he concedes, Kentucky's program can be regarded as only a partial success, even by the most generous standards. Speculating about its uneven results, he lays much of the blame for its failure on flagging federal interest and quixotic national support.

Ernest J. Yanarella and Herbert G. Reid place a local environmental dispute into the context of Kentucky's synthetic fuels push and the national campaign to develop and commercialize synfuels technology. In "Class-based Environmentalism in a Small Town," they provide a case study of the controversy surrounding the rise of public opposition, organized under the banner "Concerned Citizens of Scott County," in its successful effort to overturn a joint federal-state decision to construct a multimillion-dollar, low-BTU coal gasification plant and industrial park in Georgetown, Kentucky. Drawing upon an extensive body of letters, memoranda, and available documents in the public domain, as well as a series of in-depth interviews of the principals in the debate, the authors examine the factors that precipitated the growth of local opposition and led to the withdrawal of the industrial partner in this coal gasification venture.

Yanarella and Reid identify four factors that influenced the success of the citizens group in defeating the Georgetown project. First, they observe that once environmental opposition arose, ERDA refused to enter negotiations with the industrial developer, Irvin Industrial Development, Inc., until the environmental issues surrounding the project were addressed in a manner satisfactory to the citizens group. Second, they argue, Irvin Industrial Development carried the burden of the legacy of Garvice Kincaid's aggressive business style and questionable past dealings with the Georgetown community and then compounded that negative style by trying to win over a community with a strong controlled-growth history and zoning tradition by using a land developer's sales pitch that fell on deaf ears among the upper middle class. Third, they show that, given the interpenetration of state administration and the corporate power of the coal industry, the Kentucky Energy Cabinet could not assume a position independent of the industrial developer, and thus the state energy agency eroded public confidence in its capacity both to develop synfuels and to protect the environment and the public health. Finally, they attribute the victory of the citizens group to their ability to fashion a political strategy that at each turn in the dispute was able to use its considerable political resources to enlist wider elements of public support within the community, to enlarge the arena of political conflict, and ultimately to outlast the patience and will of Irvin Industrial Development. In sum, the authors find that the Georgetown project proved to be a harbinger of the failure of the federal government's effort to commercialize coal gasification technology.

In "Public Ambivalence about Synthetic Fuels and Other New Energy Development," Cynthia Duncan and Ann Tickamyer investigate the evanescent phenomenon of public opinion. They find that despite a growing body of survey research, social scientists have had little success in interpreting public opinion on new energy development projects like synthetic fuels plants. Their study examines the reasons for this limited understanding by reevaluating a number of representative studies conducted in western energy development areas and then comparing these findings with their original research in Kentucky. The western public opinion studies conclude that respondents typically favor development projects because of their anticipation of economic benefits. Duncan and Tickamyer were skeptical about the finding that little variation existed across subgroups because many survey instruments failed to determine whether people might, as the authors believe, simultaneously perceive economic development as beneficial and costly. This hypothesis led them to survey several Kentucky counties. Their public opinion findings suggest that public attitudes toward large development projects like synthetic fuels facilities are complex and reflect a discernible public equivocation about how to weigh energy-development costs and benefits. Their study also suggests that growth-machine theories of industrial boosterism require modification in the case of energy development in order to take into account growing community concern about the allocation of social and environmental costs stemming from the impacts and disruptions of synfuels plants and other high-technology energy projects.

In "Socioeconomic Impacts of Large-Scale Development Projects in the Western United States," Larry Leistritz and Steve Murdock draw upon the vast literature on local socioeconomic impacts of large-scale resource development projects to infer the likely character of such consequences for western communities where many synthetic fuels plants were to be located. They explore the relevant literature on the use of impact management strategies to alleviate socioeconomic impacts, because so few commercial-scale synfuels facilities reached the construction stage. They find that, despite the generally positive economic impacts anticipated, the actual impacts that various studies have uncovered suggest a mixed balance sheet. This research reveals, for example, that these large-scale resource-development projects typically draw the overwhelming majority of their construction work force from the nonlocal labor pool. Higher employment from these projects, they point out, also increases labor competition, escalates labor costs, and increases worker turnover for local firms. These studies also indicate the difficulty in measuring the secondary employment effects and the benefits derived by local communities from secondary and tertiary industries locating around major projects.

The research on the demographic impacts of large-scale projects, such as synfuels plants, has tended to focus upon the boomtown phenomenon. Here, again, Leistritz and Murdock find that these studies disclose a multiplicity of factors that complicate any easy projection of rapid population growth in impacted communities. Public service and fiscal impacts are most likely to be severe in housing, recreation, and social services, but these projects and their cancellation or closure also have an impact on the social structure of rural communities: the crime rate, juvenile delinquency, and marital problems. The authors then turn to a review of the techniques devised to control the potential community impacts, the means of minimizing demands on local systems, and steps most likely to enhance the capacity of local communities to achieve successful impact management. While cognizant of the wide range of potential impacts flowing from large-scale development projects and the local communities' pervasive fear of devastation, the authors conclude that timely planning and adequate financial resources can do much to mitigate and manage the full panoply of impacts attendant from such large-scale projects.

In Part III of this volume two essays examine the comparative contours and future prospects for synfuels policy. In "Synthetic Fuels Abroad: Energy Development in High Energy Dependency Areas," Joseph Rudolph focuses his attention on the rise and decline of the synfuels alternative in Great Britain, West Germany, Japan, and South Africa. This comparative study is highly instructive because it demonstrates that even countries with fewer native energy resources and a greater level of energy dependence than the United States have fared no better in developing synthetic alternatives to imported oil and natural gas.

The obstacles to synthetic fuels programs in these energy-dependent countries, Rudolph argues, were, in part, unique: their different institutional settings and cultural foundations. On balance, however, indigenous factors had less of an impact on the fate of these synthetic fuels research, development, and commercialization programs than the pluralistic character of decision making in western democracies, the financial problems of subsidizing capital-intensive projects in a recessionary period, the continued technological immaturity of synfuels technologies, the rightward shift in regime changes in oil-importing nations during the past ten years, the altered landscape of the world oil market in the eighties, and the misplaced emphasis on commercialization in these programs after the second oil crisis in 1979. These general factors are illustrated in a case study of the British Point of Ayr project. Then Rudolph turns to South Africa to examine the exceptional nature of its SASOL projects.

Cooperation among corporations within nations in advancing synfuels commercialization has been difficult, and so has collaboration among nations and international organizations. Whether in the European Communities or the International Energy Agency, Rudolph demonstrates that entropic tendencies within the international system have prevailed over halting efforts to generate bilateral or international support and cooperation. Since energy security is a high national goal during periods of crisis, no country seems willing to trust the matter to a multinational action. Nor do nations appear willing to sustain even fledgling national programs when an energy crisis abates. This sober comparative evaluation leads Rudolph to conclude that a realistic, long-term synfuels program will have to await the crystallization of a number of fortuitous factors, the most important of which is the stimulation of the political will to commit a nation to a long-term program without regard to energy's constantly changing international status.

Thomas Wilbanks' "Prospects of Synthetic Fuels in the United States" brings this volume to a close. Synthetic fuels development, he observes, faces an uncertain future even though oil and natural gas will eventually be too expensive for widespread use. The depletion of these finite resources, he believes, will undoubtedly spur renewed interest in synfuels, since end users, accustomed to hydrocarbon fuels, will find it easier to switch to synthetic forms rather than other energy alternatives. These uncertainties do not, however, erase the many imponderables influencing the renewal of technological development and commercialization of synthetic fuels in the short and middle terms. Uncertainties about the extent of global petroleum and natural gas reserves, the potential for breakthroughs in other energy technologies, and the availability of public funds for subsidizing the synfuels option make policies supportive of synthetic fuels development extremely problematic.

Echoing Crow's earlier analysis, Wilbanks enumerates the prerequisites for a firmly established and uninterrupted synthetic fuels research program and industry over the next twenty years, which, given the prevailing political and economic circumstances, he believes would be entirely fortuitous. A more likely scenario is that the nearly moribund and increasingly scattered American synfuels program will remain stalled for the rest of the decade due to scarce public capital and unfavorable market conditions. Perhaps by the nineties, energy shortfalls or supply disruptions will trigger anew calls for accelerated synthetic fuels development. Wilbanks believes that a place for synfuels pursued intelligently and without interruption may not be found on the nation's public agenda until the early part of the next century. At that time, a genuine commitment to the synfuels option may be possible given a clearer outline of the energy needs and technological options and the existence of creative political leadership.

Acknowledgments

It is always a happy exercise to acknowledge the many people whose thought, energy, and care have contributed to the writing of a book. We are deeply grateful for the insight, effort, and patience of the contributors of this volume. Their willingness to accept with good cheer our editorial suggestions and revisions of their early drafts has made our work much less mentally taxing and emotionally stressful. We would also like to express our gratitude to Kim Hayden at the University of Kentucky and the staff people at other institutions who typed each draft with speed and accuracy. While it is a routine convention of authors to express a debt of gratitude to their spouses, it is for us a necessary and deserved one. The critical eyes, tolerant ears, and warm hearts of Elizabeth Yanarella and Rowena Green made the editing of this work much easier and more enjoyable. We would be negligent if we did not acknowledge the role of Mildred Vasan as Greenwood Press's law and political science editor in this undertaking. Her enthusiastic reception of our book proposal, her frequent shows of quiet competence, and her guiding hand in its preparation delighted us. We are also grateful to Lyle Sendlein for his encouragement of the project and his willingness to underwrite the costs of production of the tables and figures. Finally, in dedicating this book to our parents, we acknowledge in a small way their involvement in our personal and intellectual development. They were our first teachers. We hope they find in this book some sign of the lessons we learned from them.

Part I

HISTORICAL PERSPECTIVES

1

Business, Government, and Markets: Synthetic Fuels Policy in America

RICHARD H. K. VIETOR

"The United States," reported the *New York Times*, "is on the threshold of a profound chemical revolution. The next ten years will see the rise of a massive new industry which will free us from dependence on foreign sources of oil. Gasoline will be produced from coal, air, and water" (1948). This prediction, which could easily have been made in 1980 when Congress created the United States Synthetic Fuels Corporation, was actually published in 1948, a year before the United States became a net importer of petroleum. It was based on the prospect of domestic fuel shortages and a successful outcome in the Truman administration's synthetic fuels program.

This chapter describes a succession of federal programs, from 1944 to 1986, that were supposed to encourage commercial development of synthetic fuels. Each of these programs failed either to sustain itself or to achieve its goals. This essay explains those failures in terms of (1) changes in energy markets, (2) the relationship between markets and politics, and (3) institutional channels for business-government relations.

Since World War II, energy markets in the United States, particularly for liquid fuels, have experienced four dramatic shifts in the balance between supply and demand; equilibrium has been fleeting at best. The first period, from 1942 through 1948, was dominated by shortages and rising prices. Wartime demand, limitations on imports, and lagging development of petroleum production capacity were responsible. For liquid fuels (gasoline and heating oil) these conditions actually worsened in the immediate postwar boom, precipitating the first "energy crisis" in 1948.

In the second period, from 1949 to 1968, energy shortages gave way to surplus, putting downward pressure on price. By 1959 the flood of cheap oil from the Middle East, on top of spare domestic capacity for coal, petroleum, and natural gas, caused enough market instability that the Eisenhower administration succumbed to pressures for import quotas on oil. For the next nine years these quotas succeeded in stabilizing nominal prices while discouraging any new domestic capacity.

The third period, a time of spot shortages and rising prices for petroleum and natural gas, commenced in 1969 after two decades of falling real (inflation-adjusted) prices (Vietor, 1984, chaps. 3, 8, and 14). By 1972, with spare domestic production capacity exhausted, market power passed to the Organization of Petroleum Exporting Countries (OPEC). Nine years and two energy crises later the price of oil was 1,700 percent higher at $36 per barrel. Natural gas prices had risen nearly as much, and coal prices had quadrupled.

In the spring of 1981 perceived shortages and rising prices once again gave way to surplus. Slower economic growth, the effects of conservation, and new supplies throughout the world had undermined the cartel's power. For the next four years oil and gas prices slipped lower as OPEC struggled to enforce production cuts. Then in 1985 when Saudi Arabia increased output world oil prices tumbled from $26 to $9 a barrel.

These disequilibria in energy markets gave rise to political pressures for and against the government's involvement in synthetic fuels. During periods of shortage or perceived shortage, when the costs of market failure were diffuse, the political process resembled various "public-interest" theories; bureaucratic and congressional interests labored to devise programs that would mitigate the problem (Baumol, 1965; Olson, 1965). But during periods of surplus, when prices fell and the costs of market failure were more focused, the political process is better explained by "private-interest" theories; firms and industry groups contended to use government programs to their own best advantage (Nash, 1968; Kerr, 1968; for theories of capture see Kolko, 1963; Stigler, 1971, pp. 3–21). And in both situations, institutional and informational constraints, together with the behavioral characteristics of bureaucracies, help explain the results of these public policies (Anderson, 1981; Porter and Sagansky, 1976, pp. 263–307).

Since World War II, four major public initiatives to develop synthetic fuels from coal and oil shale have come and gone: Harold Ickes's liquid fuels program (1944–1953); the Office of Coal Research pilot-plant program (1960–1968); the Energy Independence attempts at commercial demonstration (1974–1979); and the Carter administration's Energy Security program (1980–1986). The market circumstances described above first motivated and then undermined each attempt. In the remainder of this paper we will explore the history of these programs, especially the interaction of markets and institutions that explains this persistent record of policy failure.

THE SYNFUELS POTENTIAL

Synthetic fuel generally refers to artificial petroleum or natural gas chemically derived from solid fuels such as bituminous coal or oil shale. The U.S.

Geological Survey has estimated that recoverable reserves of bituminous coal and lignite in the United States amount to 1.2 trillion tons. Of this total, at least 400 billion tons is thought to be commercially feasible. Annual production currently runs at about 800 million tons. Oil shale holds even greater promise. Low-grade deposits of this fine-textured, sedimentary rock containing "kerogen" occur in 30 states. The world's richest concentration, yielding at least fifteen gallons of liquid fuel per ton, is the Green River Formation, covering 17,000 square miles where Colorado, Utah, and Wyoming intersect. These deposits contain an estimated 1.2 trillion barrels of liquid, of which 480 billion barrels are considered technologically accessible (U.S. Department of the Interior, 1968, p. 15).

Synthetic fuels have a fairly long history. In Europe as early as the 1760s and in America before the Civil War, gas was produced from coal and kerogen from oil shale. In the latter part of the nineteenth century, town gas from coal was widely used in urban areas for lighting. By the 1920s, though, the development of pipelines facilitated its rapid replacement by much cheaper natural gas, an abundant by-product of oil production. Meanwhile, during World War I, President Woodrow Wilson created Naval Oil Shale Reserves on 130,000 acres of the richest Green River deposits to foreclose further claims by placer miners and speculators. Commercial interest in oil shale continued through the 1920s, with the Bureau of Mines doing some bench-scale research and Union Oil of California operating a pilot plant (for a general history of oil shale see Welles, 1970).

In America all such commercial activities ended abruptly in 1932 when the Great Depression and an oil glut drove petroleum prices to ten cents a barrel. But in Germany, where scientists had earlier developed sophisticated processes for synthesizing both oil and gas from coal, Hitler's autarky program encouraged commercial-scale development. By 1942 Germany was synthesizing about half of its gasoline, diesel oil, and aviation fuel (Kramer, 1978, pp. 394–422; Hughes, 1969, pp. 106–32).

SHORTAGES AND THE NATIONAL SECURITY CLIMATE

The experience of mechanized war between 1941 and 1945 convinced American policy makers that liquid fuels were critical to the national security. Yet petroleum was clearly a wasting assset. In January 1943 Michael Straus, director of the War Resources Council, initiated the first synfuels program. In a memorandum to his boss, Harold Ickes, Straus complained about the "piddling" research budget of the Bureau of Mines for laboratory work on coal hydrogenation. He suggested instead that the government skip the pilot-plant stage of development altogether: "I recommend we raise our sights and dramatize this possibility by the Department's stepping right out and asking authorization and appropriation legislation to build a small commercial-size plant, possibly $20,000,000, for obtaining gasolines, kerosenes, and oils from coal." Straus appealed to Ickes's penchant for expanding his bureaucratic authority. "As supervisor of the Bureau of Mines, Solid Fuels Coordinator, Petroleum Administrator for War, Secretary of the Interior, not to mention 'Spark Plug', " he told Ickes, this was an opportunity "to surpass your splendid record for public interest and foresight" (Straus, January 28, 1943).

With Ickes's approval, Straus began building support (Ickes, January 29, 1943). To alleviate potential opposition from the oil industry, he worked with Ralph Davies, the deputy petroleum administrator for war, who had close ties with industry leaders (Straus, February 1, 1943; Davies, August 1943). In the Senate, Straus enlisted Joseph O'Mahoney of Wyoming, chairman of the Public Lands Subcommittee. O'Mahoney organized hearings and introduced legislation to provide $30 million for synfuels research. The bill authorized the Bureau of Mines to build demonstration plants that would determine the cost of synthetic fuels and encourage "expansion of a process to full commercial scale." Straus tapped Jennings Randolph of West Virginia, chairman of the Appropriations Subcommittee for Mining, to cosponsor the bill in the House (Straus, April 3, 1943, August 20, 1943, and January 27, 1944). In April 1944 the Synthetic Liquid Fuels Act passed the Senate unanimously and the House by a wide margin.

Interior officials grossly underestimated the difficulties and cost of starting a new energy industry. For the next several years, technical and legal problems plagued the program. The Bureau of Mines had to rely on German technology, German technical documents, and even German equipment.[1] Procurement offices in Washington had to approve scarce construction materials. Engineering research and plant preparation was divided into three parts: coal hydrogenation, gas synthesis, and the mining and processing of oil shale. At the naval oil-shale reserve near Rifle, Colorado, the bureau started building a demonstration plant for extracting and refining kerogen. In the town of Louisiana, Missouri, the bureau acquired an army surplus ammonia plant as the site for two demonstration units: one for direct liquefaction through coal hydrogenation and one for coal gasification by the Fischer-Tropsch process. These two units, designed for capacities of 200 and 100 barrels per day respectively, would be larger than anything operated outside of Germany (U.S. Bureau of Mines, 1949b; Boyd, 1948, pp. 25–65).

While the program moved slowly ahead, conditions in domestic energy markets actually worsened after World War II. In January 1947 a special congressional committee on petroleum, chaired by Senator O'Mahoney, recommended "bold steps" toward synthetic-fuels production. But, cautioned the committee report, "Until such time as synthetic production costs are not in excess of those of production from crude oil, there will be a natural temptation in peacetime to meet domestic deficiencies with imported petroleum" (Senate Special Committee Investigating Petroleum Resources, 1947, pp. 15, 30–31). This phenomenon, the market, would continue to undermine government policy on synfuels for the next forty years.

The heating oil shortage and voluntary allocation program in the winter of 1947–1948 induced an atmosphere of crisis. In January James V. Forrestal, Truman's new secretary of defense, recommended that Congress adopt an $8 billion synthetic-fuel program. Julius Krug, secretary of the interior, called for a government-sponsored, ten-year program to build and operate three prototype plants capable of producing 30,000 barrels per day of gasoline. In Congress bills were introduced to authorize funding for one or more full-sized liquefaction plants (*National Petroleum News*, January 21, 1948; Patterson, July 2, 1947; House Commerce Committee, 1948).

Heretofore the petroleum industry had been cautiously optimistic regarding the feasibility and prospective need for synthetic fuels. Public and private estimates of the costs of synthetic gasoline ranged from 10 to 18 cents per gallon at a time when the wholesale refinery price of gasoline was about 12 cents per gallon ("Oil Shale and Shale Oil," April 29, 1943, pp. 73–79; Wilson, July 12, 1944; Senate Interior Committee, 1948, pp. 51–59, 238–47, 303–18). There was, however, no support for a massive governmental program. Several oil-industry executives vocally opposed additional legislation. So did Bruce K. Brown, a vice president of Standard Oil of Indiana and chairman of the Military Petroleum Advisory Committee.[2] Although this opposition helped defeat any new, large-scale initiatives, Congress did approve an extension of the Liquid Fuels Act in 1948, authorizing $30 million more.

ENERGY GLUT AND THE COST DISPUTE

Harry Truman had scarcely declared an energy emergency in the spring of 1948 before it was over. Once wartime price controls were lifted in 1947, oil prices more than doubled from $1.19 to $2.61 per barrel. Within a matter of months, domestic oil stocks increased and cheap foreign crude oil poured into the east coast from Venezuela and the Middle East. Coal output reached record levels, while new pipelines opened industrial markets to natural gas. When economic growth turned to recession in 1949, it became suddenly clear that the expansion of energy capacity had been excessive. Several major oil companies canceled ongoing synfuels experiments, and the federal demonstration program was exposed to more criticism than ever.

The growing reliance on imported oil gave the program a fresh rationale. At the dedication of its liquefaction units in 1949, the Bureau of Mines described them as "forerunners of a new basic industry that ultimately may free the United States from dependence on foreign liquids" (U.S. Bureau of Mines, 1950, p. 143). Although the gas-synthesis unit was not completed, the coal-hydrogenation unit had been tested successfully, and the work at Rifle was yielding good data (U.S. Bureau of Mines, 1951, pp. 17–19; Krug, February 9, 1949; Friedman, February 23, 1949; Day, February 24, 1949). Based on these initial efforts, the Bureau of Mines published detailed cost estimates for commercial scale. The report concluded that gasoline could be produced from coal for 11 cents a gallon, a competitive market price (U.S. Bureau of Mines, 1949b).

The oil industry disagreed. The bureau's report not only insinuated that oil companies themselves should be developing synthetics, but lent credence to calls for government production. To resolve what soon became an acrimonious debate, the new Interior secretary, Oscar Chapman, asked the National Petroleum Council to establish a blue-ribbon committee to review the bureau's cost estimates and provide its own independent estimates (Chapman, April 21, 1950). The council, created by President Truman in 1946 to advise the interior secretary on matters of energy policy, consisted of about 80 chief executive officers from major oil companies and the independent producers and refiners.

The Petroleum Council chose W.S.S. Rodgers of the Texas Company to head the synfuels study. His coordinating committee included Bruce Brown and Max Ball, formerly the director of the Oil & Gas Division of the Interior Department. To staff the critical subcommittee on production costs, Brown recommended several executives previously on record as opposing synfuels (Brown, June 27, 1950). By the time the study was completed in 1952, more than 49 executives, aided by 150 technical personnel, had spent $500,000 to prepare the multivolume report on raw materials, process engineering, economics, and costs.

The results proved politically devastating. Based on commercial-scale designs summarized in Table 1–1, the Petroleum Council estimated costs of 16.2 cents per gallon for gasoline from shale and 41 cents for coal hydrogenation. These estimates grossly exceeded the bureau's figures of 11.5 and 11.0 cents per gallon respectively (U.S. Bureau of Mines, October 25, 1951a; National Petroleum Council, 1951). The disputed costs fell into four categories: capital costs, operating costs, by-product credits, and finance charges. Problems of perception accounted for most of the differences. In its design, the Petroleum Council minimized by-product output; the bureau chose to maximize it. The council allowed for employee housing, but the bureau did not. Most significant, however, were finance charges. The council figured on 100 percent equity financing; the bureau assumed 60 percent debt. The discrepancies were so large that Secretary Chapman hired Ebasco Services, an engineering firm, to prepare an independent estimate (National Petroleum Council, January 29, 1952, July 29, 1952; U.S. Bureau of Mines, October 25, 1951b, pp. 11–16). The cost estimates of all three parties are compared in Table 1–2.

In February 1953, just a month after Dwight Eisenhower's inauguration, the National Petroleum Council released its final report on synfuels. After all the debate, nothing had changed: "As shown by this extensive and conclusive study," read the report, "all methods of manufacturing synthetic liquid fuels proposed by the Bureau of Mines are definitely uneconomical

Table 1–1
Summary Data from 1951 Designs for Coal-Hydrogenation and Oil-Shale Plants

	National Petroleum Council's hydrogenation plant	Bureau of Mines hydrogenation plant	Joint design for oil-shale plant
Coal or shale input (tons per day)	12,960	14,800	76,800
Products (barrels per day)			
gasoline	19,490	18,600	25,380
liquefied petroleum gas	6,390	7,100	1,780
other fuels	---	---	12,540
total liquid fuels	25,880	25,700	39,700
Chemicals (barrels per day)	1,120	3,960	---
Ammonia (tons per day)	---	359	92
Sulfur (tons per day)	---	47	43
Construction materials (tons of steel)	220,000	217,130	178,000
Investment capital (millions of dollars)	533	400	333
Cost per gallon of gasoline with 6 percent return on equity investment (cents)	41.4	11.0	16.2 (NCP) 11.5 (BOM)

Source: Compiled from National Petroleum Council, Report of the National Petroleum Council's Committee on Synthetic Liquid Fuels Production Costs, October 31, 1951, Synthetic Fuels File, NPC/DOI, 1951.

Table 1–2
Summary of Unit-Cost Estimates for Producing Gasoline by Coal Hydrogenation

	National Petroleum Council (10/31/51)	Bureau of Mines (10/25/51)	Ebasco Services (March 1952)
Operating costs	25.3	17.7	19.5
Housing costs	2.6	---	---
Finance charges	19.0	8.2	22.0
Total process costs	46.9	25.9	41.5
Less by-product revenue	-5.5	-14.9	-13.4
Total costs	41.4	11.0	28.1

Note: **Cents per gallon**

Source: National Petroleum Council, "Interim Report of the NPC's Committee on Syntehtic Liquid Fuels Production Costs," July 25, 1952, file 106.61, NPC/DOI, 1952, p. 8.

under present conditions." Since "the need for a synthetic liquid fuel industry in this country is still in the distant future," and "since new techniques may be available then," the Petroleum Council "question[ed] the wisdom of the Government financing large-scale demonstration" (National Petroleum Council, February 26, 1953, pp. 10–11).

When the Republican-dominated House Appropriations Committee opened hearings on the budget late in March, no one in the new administration wanted to continue the synfuels demonstration. After Interior Secretary Douglas McKay recommended cutting the program's funds, the House subcommittee terminated them altogether. Democrats remonstrated loudly, but to no avail. Representative Emanuel Celler accused big oil companies of "scuttling the plan," while Congressman Carl Perkins charged that the process had "proved successful and has reached the point of being competitive with crude oil. Yet, because of that fact, we want to destroy the process in favor of the oil lobby." Perkins pinned the demise of synfuels squarely on the Petroleum Council (House Appropriations Committee, March 11, 1953, pp. 627–28, 1006; *Congressional Record*, 1953, pp. 3355, 4022–26, 4120, and 4145). Early in 1954, the government shut down the plants at Louisiana. After some further debate in 1955, funds were cut off for oil shale too. The facility at Rifle was mothballed in 1956 (National Petroleum Council, January 25, 1955, and May 5, 1955).

RESEARCH: THE CURE FOR A "SICK INDUSTRY"

Coal was a "sick industry" in more ways than one by 1956. Coal production had declined 39 percent from its peak of 630 million tons in 1947, three thousand mines had closed, and the number of mine workers had fallen by nearly half. Coal's share of the energy market had fallen by 18 percent with no end in sight. Spokesmen for the coal industry attributed their problems more to government policies than competition: low natural gas prices, tax advantages for oil producers, the absence of quotas or significant tariffs on imported oil, and especially the Atomic Energy Commission's massive promotion of nuclear electric power. According to the chief lobbyist for the National Coal Association, "the AEC would spend more in 1957 than the entire profit of the coal industry since 1920" (House Interior Special Subcommittee on Coal Research, 1957a, pp. 52–56, and 1957b, pp. 556–67). This seemed unfair.

But structural fragmentation, not government policy, was the real problem. With few apparent economies of scale, no significant integration had occurred in the coal industry. Of an estimated 4,000 concerns, less than 1,000 reported net income. In the early 1950s, profits averaged $22,000. Since structural weakness was relatively intractable politically, the industry's leadership focused on research. "Unbelievable as it may seem," said one coal executive, "the amount spent for research by some of the industrial groups is about

10 times larger than the total amount of net profit made by our entire industry." Indeed, of the $17 million spent on coal research in 1955, coal producers contributed only $2.4 million. The rest came from government sources, equipment manufacturers, and industrial consumers (House Interior Special Subcommittee on Coal Research, 1957a, pp. 34–35, and 1957b, p. 38).

Not surprisingly, coal-state congressmen, the mine workers, and some of the industry's leaders desperately grasped at federally sponsored research in synthetic fuels as a means of relief. But the industry's political effectiveness suffered from the same fragmentation that hurt profitability. Thus, in 1954, when a group of coal executives approached the Eisenhower administration for help, they received consolation and an interdepartmental study, but nothing of substance (Cooper, May 21, 1954; Report of the Interdepartmental Committee, October 7, 1960). In 1956, a strenuous effort by Representative John Saylor of Pennsylvania achieved no more than a resolution to investigate R&D as a means of stimulating "economic revival in the coal industry" (House Resolution 400, 1956; House Interior Special Subcommittee on Coal Research, 1957b, p. 1). At hearings on this subject in 1956 and 1957, witnesses testified to the industry's woeful condition, citing inadequate research as a critical problem. Yet spokesmen for the administration were loathe to expand the government's role, especially after the previous effort "had become too large a project." Beyond the pilot-plant stage, said one Interior official, "it was desirable for the Government to step out" (House Interior Special Subcommittee on Coal Research, 1957b, pp. 10, 16).

Spokesmen for the coal industry proposed a Coal Research Foundation with a fifteen-member board of directors to dispense up to $50 million a year in federal funds for coal research. The board would be composed of experts from government, academia, and, mostly, industry. In 1957, the House subcommittee recommended an independent Coal Research and Development Commission, modeled along the lines of the Atomic Energy Commission. Such a commission, with advisory committees representing the coal industry and coal users, would "not be shackled and inhibited by such traditional approaches and restrictive policies as control the research activities of the Department of the Interior." With an initial appropriation of $2 million, the commission would pursue "promising short-range possibilities." Coal-derived synthetic fuels were not mentioned (House Interior Special Subcommittee on Coal Research, 1957b, pp. 548–49, and 1957a, p. 91).

Several bills to create such a commission were introduced in 1958. The administration proposed instead an Office of Coal Research within the Interior Department, making research contracts only with nonprofit institutions (Hardy, March 4, 1959, p. 13). Although the Senate passed an independent-commission bill, Republicans in the House managed to kill it. In 1959 Wayne Aspinall, a Colorado Democrat, succeeded to the chairmanship of the House Interior Committee. Aspinall reintroduced the independent-commission bill, but the administration still opposed it. This time both houses passed coal-research legislation that President Eisenhower vetoed on the grounds that an additional agency would dilute the Interior Department's "established interest" (Senate Interior Subcommittee on Minerals, Materials, and Fuels, 1959; Eisenhower, September 16, 1959).

Left with no alternative, House Democrats introduced compromise legislation in 1960 to establish an Office of Coal Research (OCR) within the Department of the Interior. With an initial appropriation of just $1 million, the bill easily passed both houses and was signed by the president (Hardy, December 29, 1959; House Interior Committee, 1960). The Coal Research and Development Act of 1960 put the federal government back in the business of cooperative coal research; but at what level, in what directions, and to what ends was uncertain. Significant progress would run squarely in the face of the oil glut.

THE SECOND SYNFUELS PROGRAM

The government's new program for coal research got off to a very slow start. After a year Wayne Aspinall complained that it was "similar to the history of a patient who died before the doctor, who lived in a building adjacent to him, knew where he was" (House Interior Committee, 1961, p. 171). Under pressure, the Interior secretary finally appointed George Lamb from Consolidation Coal as director of the Office of Coal Research and a panel of researchers and executives from coal companies, utilities, railroads, equipment manufacturers, and the United Mine Workers to serve as a General Technical Advisory Committee. In June of 1961 the committee met to adopt procedures and develop a research strategy. It outlined seven project areas, of which one involved synthetic fuels. But beyond "short-range research systems" there was no clear objective (General Technical Advisory Committee, June 6, 1961, p. 2).

In a matter of months, a number of high-technology coal-conversion projects had attracted the OCR's attention as well as the interest of congressmen responsible for the OCR's appropriations. While several small studies were under consideration by the divisions of Marketing and Mining, the OCR's Division of Utilization had identified five synthetic-fuel projects (Office of Coal Research, February 18, 1962). Once real projects were on the table, a dispute developed among OCR staff and members of the Advisory Committee over whether the OCR should commit resources to synthetic-fuel pilot plants or shorter-term projects of immediate benefit to coal producers (Charmbury, December 13, 1961; Lehner, December 13, 1961; Sporn, December 2, 1961; Elrond, December 19, 1961).

The issue was eventually decided by the power of the pork barrel. Senator Robert Byrd of West Virginia, a member of the Senate Appropriations Committee and stalwart advocate of coal research, moved quickly in 1962 to assert

Table 1–3
Office of Coal Research Synthetic Fuels Pilot Plants, 1962–1972

Project	Prime contractor	Contract duration	Government funding level	Location
Project Gasoline	Consolidation Coal Company (a)	8/63 - 3/72	$20,377,000	Cresap, WV
Project COED	FMC Corporation	5/62 - 2/74	19,332,000	Princeton, NJ
CO2 Acceptor	Consolidation Coal Company	6/64 - 9/73	16,606,000	Rapid City, SD
HY-GAS	Institute of Gas Technology	7/64 - 7/72	14,399,000 2,384,000 (d)	Chicago, IL
SRC				
Low Ash Coal	Spencer Chemical Company (b)	8/62 - 2/65	1,240,000	
Solvent Refined Coal	Pittsburg & Midway Coal (b)	10/66 - 4/72	7,640,000	Ft. Lewis, WA
BI-GAS	Bituminous Coal Reasearch, Inc. (c)	12/63 - 2/71	3,438.000	Homer City, PA

[a] Consolidation Coal was aquired by Continental Oil Co. in 1966.

[b] The Solvent Refined Coal Project started out as Low Ash Coal. Pittsburg & Midway Coal was a subsidiary of Spencer Chemical. In 1963, both firms were aquired by Gulf Oil Corp.

[c] BCR, Inc. is the research sudsidiary of the National Coal Association.

[d] American Gas Association contribution.

Source: Office of Coal Research, Annual Report, 1967 and 1972.

his influence. By lining up support from the White House and the United Mine Workers, he succeeded in boosting the OCR's budget to $3.5 million, mostly earmarked for two pilot-plant projects in West Virginia. Members of the General Technical Advisory Committee scarcely noticed, much less balked at, the program's change of course.

The OCR's commitment to synthetic fuels grew steadily, albeit slowly, over the next five years. Budgetary authority remained a bottleneck. Although appropriations increased 45 percent annually, they had scarcely reached $14 million by 1969. Since the costs of pilot-plant projects were invariably underestimated, OCR staff members were constantly making trade-offs between starting new projects and nurturing those in progress. By 1968, the Division of Utilization was managing $53 million in contracts for synfuel pilot plants; the Marketing and Mining divisions shared $4.2 million in contracts (House Science and Astronautics Subcommittee on Science, Research, and Development, 1969, p. 306; Office of Coal Research, 1969, pp. 22–25).

Six pilot plants eventually comprised the heart of the OCR's program to develop synthetic fuels (see Table 1–3). Three each were devoted to coal gasification and coal liquefaction. The projects ranged in size from the 75-ton-per-day HY-GAS plant near Chicago to the 100-pound-per-hour gasifier developed by Bituminous Coal Research, Inc., in Pennsylvania. The six plants utilized substantially different technologies.

Of the three liquefaction technologies, Solvent Refined Coal was the least complicated. Its purpose was to remove contaminants from coal to facilitate cleaner combustion. In the SRC process, crushed coal was dissolved in a solvent and then filtered to remove the ash. With further treatment it could be liquefied, although that was not the thrust of the OCR contract. Project COED (Char Oil Energy Development) was similar to petroleum refining; fractions of liquids and gases were released from fluidized beds at successively higher temperatures. Project Gasoline was the most complicated of the projects politically as well as technically. From coal to synthetic crude oil, this process required eleven steps; the most critical of these was direct hydrogenation at very high temperature and pressure (Office of Coal Research, 1973, pp. 45–50, and 1968, pp. 15–16). The three gasification plants, while chemically complex, did not pose such severe problems of materials handling. The CO_2 Acceptor Process was perhaps the most innovative, since it worked without either hydrogen or oxygen and was designed to treat lignite.

Selection, funding, and implementation of these large projects remained extraordinarily politicized and inept. Pork-barrel politics dominated project choice and site selection. The General Technical Advisory Committee never developed any control over program planning, much less influence in Washington. And the Office of Coal Research, as a bureaucratic entity, failed to develop strong relationships within the Department of the Interior or with any congressional committees.

Congressional "champions" proved essential for funding the pilot plants (Reichl, 1980). Among the more visible were Representative Julia Hansen and Senator Henry Jackson of Washington for the SRC Project (Tacoma), Senator Karl Mundt of South Dakota for Project Lignite (CO_2 Acceptor Process), and Senator Robert Byrd for Project Gasoline. In 1965, the SRC pilot plant received funding even though a majority of the Technical Advisory Committee initially opposed it (General Technical Advisory Committee, January 27, 1965, pp. 36–39, and July 13, 1965, pp. 37–38). Project Lignite received its initial funding from an appropriations amendment added by Mundt in 1963 even before the OCR had a technology in mind (Larson, July 2, 1963; Senate Appropriations Committee, Subcommittee on Interior Department Appropriations, February 24, 1964, pp. 659–60). And throughout the 1960s Project Gasoline (at Cresap, West Virginia) survived intense criticism and absorbed more and more funding thanks to Senator Byrd's tireless efforts on the Senate Appropriations Committee (Minutes of Meeting with Senator Robert Byrd, March 21, 1969; General Technical Advisory Committee, March 21, 1969, and May 21, 1969; Dole, May 21, 1969).

In its advisory capacity to the OCR, the General Technical Advisory Committee was virtually a rubber stamp. This was due in part to its limited charter. Its members had no staff and met only in formal quarterly sessions. Unlike the National Petroleum Council, the Advisory Committee conducted no studies of its own, relying instead on presentations by OCR staff. And the OCR did not really want independent oversight. When members of the committee did occasionally question the progress of a project, the director cautioned them "to remember that everyone of these pilot plant programs has been approved by you before. If we come around and get too many negative points of view," he warned, "we might get into a little more trouble." Not once in ten years did the OCR ever reject or terminate a project based on the committee's recommendation (General Technical Advisory Committee, January 27, 1965b, p. 9, and October 3, 1966, p. 43).

Since the Office of Coal Research was dependent on individual champions, without strong support from industry, it failed to develop the kind of bureaucratic linkages it needed to become a major federal program. To get anywhere, the OCR was continually running what George Fumich (its second director) called an "obstacle course" of four congressional committees and two executive agencies, none of which became institutional sponsors for the program until after the energy crisis. This situation, according to John O'Leary, an experienced energy bureaucrat, "cut into ribbons any effort on the part of the Government to develop the sound comprehensive balance in energy R&D policies." By contrast, in the relationship between Congress and the Atomic Energy Commission, "you find a sort of mutual reinforcement society." But in coal, said O'Leary, "we have hostile appropriations committees, and we have hostile substantive committees" (General Technical Advisory Committee, October 7, 1969, pp. 85–86).

Project Gasoline's problems epitomized the collective failings of the OCR's synfuels program during this period of energy surplus. The plant at Cresap, which was managed by Consolidation Coal, was designed as an integrated system to produce synthetic crude oil from coal. The Office of Coal Research seriously underestimated the project's capital costs, construction schedule and shakedown time, mechanical problems, and the unit costs of its end-product. In 1963 the original contract put the cost at $9.9 million; construction was scheduled to take three years (*Review of the Coal Research Program*, n.d.). As it turned out, construction took four years, shakedown operations lasted three more years, and when the project was shut down in 1970 for lack of results, $20.4 million had been expended by the government.

Mechanical, design, and maintenance problems seemed to get worse rather than better as shakedown operations continued. Filters, compressors, rotating equipment, and other critical parts kept breaking down, even with continuous modification and frequent replacement. Problems with the carbonization and hydrogenation vessels caused periodic shutdowns throughout the entire operating period. Worse still, when one section was up and running, two or three others were invariably shut down for repair. Thus, the coal liquefaction system never did operate as an integrated whole (*Brief History of Cresap, West Virginia Pilot Plant Operations*, November 9, 1971; *Monthly Operations Log*, n.d.).

When OCR officials did not get a fully operational plant for the costs they had estimated, they were quick to place blame on the contractor. Outside evaluations were ordered, the Interior Department was brought in, and eventually the contractor's parent firm, Continental Oil, was urged to remedy the problems. When nothing seemed to help, all parties involved hoped to allay congressional criticism by quietly accepting Project Gasoline's termination in April 1970 (Chief, Division of Contracts and Administration, November 27, 1967; Fumich, December 11, 1967; U.S. Department of the Interior, April 21, 1969).

STEWART UDALL'S OIL SHALE INITIATIVE

Conservation, rather than the desperation of a mature industry, apparently provided the impetus to a second oil-shale program. Quite unexpectedly in November 1963 Interior Secretary Stewart Udall solicited public suggestions for "the orderly conservation and development of the federally-owned oil-shale deposits in Colorado, Utah, and Wyoming" (Kelly, April 30, 1964). Congressional pressures, continuing controversy over unsettled mining claims, and the pending lease of the Rifle facility to a pair of oil companies may have helped put shale on the secretary's agenda. At any rate, it made sense to Udall, whose conservation ethic valued efficient utilization of resources as much as protection and had little concern for market forces (Udall, 1963). "If this resource is to make an optimum long-term contribution to the economic well-being of the Nation," said Udall in 1964, "the major public policy questions needed to be identified and evaluated at the outset" (Oil Shale Advisory Board, February 1, 1965).

The technical, economic, and political problems associated with any development of oil shale were legion. The most pressing of these was legal clarification of the government's title to the shale properties. Thirty-six thousand unpatented claims from the 1920s had never been cleared, and 16,000 newer claims (for metalliferous minerals interspersed in the shale) were still pending. Until these claims were resolved, the Interior Department could not begin "blocking-up" commercial-sized properties from the patchwork pattern of public and private ownership (*Ickes v. Virginia-Colorado Development Co.*, 1935; Union Oil Co. of California, 1964; U.S. Department of the Interior, 1968a, pp. 28–31, 38).

The shale-oil content and value of federal properties was nearly as clouded as their title. Prior to leasing, the government needed to know how rich its properties were and how much was technically and economically recoverable from any given property. An immense job of measurement faced the Bureau of Mines, the Bureau of Land Management, and the Geological Survey. Of course, the government did anticipate various technical problems, including water availability, mining and refining methods, spent-shale disposal, and resource conservation. But in the 1960s these issues still seemed "subordinate" to antitrust issues and the assurance of adequate government revenue (U.S. Department of the Interior, 1968a, pp. 100–107).

For the Interior Department to embrace even the most gradual utilization of oil shale, there had to be some degree of need and practicability. By the mid–1960s some energy analysts, including those at the Interior Department, felt that shale oil might again be "coming into the competitive range." Cost estimates that appeared during the 1960s indicated that shale oil could be produced at unit costs in the range of $1.46 to $2.68 per barrel, and be sold with reasonable profit for $2.00 to $3.50 per barrel (for detailed estimates of capital and unit costs see Merrow, 1978, p. 126; Steele, 1967, pp. 542–48; U.S. Department of the Interior, July 7, 1964, p. 40, and 1968a, Appendix B, pp. 1–21). In 1964, as if to confirm these estimates, The Oil Shale Corporation (TOSCO), Standard Oil of Ohio, and Cleveland-Cliffs Iron started a joint venture (Colony Development) to build and operate a 1,000-ton-per-day prototype plant (Senate Judiciary Subcommittee on Antitrust and Monopoly, April–May 1967, pp. 308–31).

To help clarify these issues, Udall called for comments from interested parties and appointed six prominent and relatively disinterested citizens to serve on an Oil Shale Advisory Board.[3] Oil companies generally agreed that "orderly development" was crucial, and that the department should start by developing a comprehensive set of leasing regulations (Shell Oil, Phillips Petroleum, Atlantic Refining, Union Oil, and TOSCO, January–April, 1964). The findings of the Advisory Board, released in 1965, were considerably less sanguine. Board members could only agree on objectives: encouraging technology, encouraging competition, establishing conservation standards, preventing speculation, and securing adequate royalties for the government. On issues of substance, the board was divided. At one extreme, two members urged that private enterprise be turned loose on the shale lands in a manner similar to development of the outer continental shelf. At the other extreme, two members feared a "giveaway" of unknown value to big oil companies; they opposed any leasing beyond contract research. The remaining two members favored restrictive leasing either for commercial or experimental work (Oil Shale Advisory Board, February 1, 1965, pp. 6–7, 16–39).

In January 1967 Udall announced a "Five Point Oil Shale Development Program" designed for minimal squawk. It offered something for everyone, but in limited amounts: (1) accelerated title clearance; (2) blocking-up of commercial tracts by exchanging public and private lands; (3) leasing of small experimental tracts while reserving larger commercial tracts until R&D had been successfully performed; (4) AEC experiments with in situ retorting by nuclear fission; and (5) government-funded research of mining and processing technologies (U.S. Department of the Interior, January 27, 1967).

Udall called for public comment on these guidelines, with specific suggestions for leasing regulations. His office was flooded with comments from irate citizens, congressmen, and all kinds of interest groups. The prospect of action sparked a raging controversy. "Giveaway" was the cry heard everywhere. A "Grabber Conspiracy" to match Teapot Dome caught the media's attention. With the press suggesting that five trillion dollars might be at stake, a lot of people concluded that the government was in cahoots with Big Oil to rape the public domain.[4] Senator Philip Hart immediately convened his Subcommittee on Antitrust and Monopoly to investigate, while Senators Jackson and Proxmire urged Udall to delay issuing leasing regulations until the Interior Committee could do likewise (Senate Judiciary Subcommittee on Antitrust and Monopoly, April–May, 1967, pp. 2–3; Udall, January 18, 1968, and October 24, 1968).

The upshot of this controversy was a set of leasing regulations that ignored most of the oil industry's pragmatic suggestions and quashed what little enthusiasm there was for serious investment in oil shale. Lease tracts were limited to experimental, not commercial, scale with high royalty payments, bonus payments despite project termination, and tougher-than-expected environmental controls. Senators Paul Douglas and William Proxmire, protectors of the "public interest," wrote Udall to congratulate him on thwarting "the predators" (Douglas, August 6, 1968; Proxmire, August 1968).

On December 20, 1968, the Interior Department held a test-lease sale, concluding five years of careful planning. When the bids were opened, the disappointment was stunning. Only three bids were submitted; two by TOSCO for $250,000 each, and a third for $625 by some anonymous entrepreneur. Secretary Udall hastily rejected all bids and left the "determination of what is in the best public interest" to "the incoming Administration and Congress" (U.S. Department of the Interior, December 27, 1968).

Speaking that same month to a group of independent oil producers, an assistant secretary of Interior, Cordell Moore, attributed the failure of synfuels to energy market conditions. Synthetic fuels were "much more vulnerable, much more sensitive to potential manipulation of oil import levels than is oil from conventional deposits. "Our oil imports policy," said Moore, "casts a long shadow over the prospects we have for bringing synthetic liquid fuels from coal and oil shale to market" (Moore, December 2, 1968).

ENERGY CRISIS AND ENERGY INDEPENDENCE

Even before the OPEC price shocks of 1973–1974, domestic shortages of natural gas and the demand for clean low-sulfur fuels reinvigorated the government's ongoing programs for developing synthetic fuels. In 1971, when President Nixon first addressed incipient energy problems, he called for "greater focus and urgency" with the synfuels projects in progress. He announced a "cooperative program with industry to expand the number of pilot plants" and "the orderly formulation of a shale oil policy—not by any headlong rush toward development, but rather by a well considered program in which both environmental protection and the recovery of a fair return to the Government are cardinal principles" (Nixon, June 4, 1971). In political terms, these measured initiatives just matched market conditions, well short of crisis.

In coal research the pace quickened, but with more careful technical planning and some substantive cooperation between business and government. After the Project Gasoline fiasco, the Office of Management and Budget issued guidelines for synfuels research, suggesting that industry contribute one-third of project costs. Indeed, the American Gas Association, whose members faced a worsening reserve situation, helped conceive this joint-venture approach. In August 1971 the Gas Association and the Office of Coal Research signed an agreement to fund six pilot plants over the next four years; a second phase, involving commercial demonstration, would follow later. The program was jointly managed by the research director of the AGA and the director of the OCR. A steering committee composed of the assistant secretary for mineral resources, the president of the National Academy of Engineering, and the president of the AGA would resolve any serious disputes (U.S. Office of Management and Budget, December 18, 1970; General Technical Advisory Committee, January 12, 1971; *Agreement between Department of the Interior and the American Gas Association*, August 3, 1971).

This approach proved far more effective than the OCR's politicized contract management of the 1960s. Projects, plant sites, and funding levels were determined according to technical and economic merit. By 1976 most of the pilot plants had operated successfully, yielding data necessary for three commercial demonstration proposals (American Gas Association, June 1973).

The Nixon administration's planning for oil-shale development proceeded similarly along a measured, albeit more controversial, course. When Walter Hickel, the new secretary of the Interior, designated an Oil Shale Task Force, western boosters urged him to take an approach more palatable to business. But organized environmentalists, eastern congressmen, and a host of other concerned citizens railed against another "giveaway." Although Hickel's task force eventually produced a draft leasing program, public pressure prevented its release (U.S. Department of the Interior, January 1969–July 1970).

In 1971 Secretary Hickel's successor, Rogers Morton, restarted the process of developing a prototype leasing program. Morton's idea was to "establish stable economic and regulatory policies," but then leave development risks to the private sector (Morton, January 10, 1973). The Interior Department carefully reviewed the 1968 program, prepared environmental and socioeconomic impact analyses, and held a series of public hearings. The program it proposed was much more sophisticated than earlier efforts. Core drillings by interested firms would lead to nominations of prospective development sites. Interior would select six tracts of 5,120 acres, two each in Colorado, Wyoming, and Utah. The lease form not only provided for up to 30 years of commercial production, but tied royalties to commercial output and eliminated punitive provisions in the event of nonproduction. After preparing final, site-specific environmental impact statements, Interior would conduct a competitive lease sale, scheduled for December 1972 (U.S. Department of the Interior, December 1973).

But environmentalists deemed the draft environmental impact statement completely inadequate.[5] Faced with the threat of litigation, the Interior Department withdrew its draft for substantial revision. Only when the final statement was released in September 1973 could Secretary Morton announce the sale, scheduled for January 8, 1974. Seven days before the sale OPEC raised the price of crude oil from $3.00 to $11.65 a barrel. The oil shock had an immediate and dramatic impact on the government's synfuels policy. The policy at issue was no longer whether government should subsidize development of synfuels, but rather how fast and in what manner.

The Office of Coal Research, a minor and somewhat tarnished bureau within the Department of the Interior, was clearly not adequate to the task. Indeed, the oil shock forced the issues in an ongoing debate between President Nixon and Congress over energy reorganization. The president had earlier proposed creation of a Department of Energy and Natural Resources to consolidate all energy functions of government. But Senator Henry Jackson, chairman of the Sentate Interior Committee, favored legislation to create five independent, government-owned energy-development corporations (Senate Interior Committee, 1973, pp. 2–60). Outside of government, environmental and antinuclear groups were urging separation of the Atomic Energy Commission's regulatory and promotional functions.

In the heat of crisis, Nixon and Jackson hurriedly compromised on the Energy Research and Development Administration (ERDA). ERDA would be a straightforward merger of the research functions of the Atomic Energy Commission, the Office of Coal Research, and the Bureau of Mines. To provide coordination and greater flexibility in energy development, ERDA would be headed by a single administrator. But to prevent domination by nuclear interests, the law designated six assistant administrators, one each for nuclear, national security, fossil fuels, solar, geothermal, and advanced power systems (Public Law 93–438, October 11, 1974; House Science and Technology Committee, 1978, pp. 51–79).

ERDA's initial appropriations for coal research were generous by historical standards, but could scarcely support large-scale commercial demonstration. In February 1976 Gerald Ford proposed a "Synthetic Fuels Commercial Demonstration Program" providing $11 billion to help industry build twelve demonstration plants. But with gasoline stocks abundant, oil prices stable, and the natural gas shortage easing, political support for that level of funding had long since evaporated. Even modified proposals, for $6 billion and then $4 billion, failed. "Is this truly to get knowledge," asked one cynical Democratic congressman, "or is it to finance the major energy companies in commercial production of these fuels?" (Ottinger, 1976).

The Carter administration, as part of its broader, conservation-oriented National Energy Plan, supported a limited effort at commercial demonstration. The Department of Energy, created by another reorganization in 1977, let several contracts for developing preliminary designs of commercial-scale gasification and liquefaction. Two of these were eventually selected for financial support on a fifty-fifty basis: the Great Plains Gasification Project, by a consortium of gas utilities, and SCR-II, a liquefaction technology owned by Gulf Oil, but originally developed as Solvent Refined Coal by Spencer Chemical and the OCR (U.S. Department of Energy, 1978, pp. 57–89; Gulf Research and Development Co., 1982).

Oil shale, too, got off to a quick start before bogging down later in the seventies. At the Prototype Lease Sale in 1974, four consortia of oil companies made acceptable bids. Rio Blanco Oil Shale, a joint venture between Gulf Oil and Standard Oil of Indiana, bid $210 million for tract C-a, the choicest lease in Colorado, containing an estimated yield of four billion barrels. C-b went for $117 million to a group headed by Atlantic Richfield. The less valuable tracts in Utah, which received bids of $175 and $45 million, were combined in a single project by Phillips, Sun, and Sohio (Senate Interior Subcommittee on Minerals, Materials, and Fuels, March 1976, pp. 32–34).

Work began promptly on the detailed development plans and baseline environmental studies that were required by the leasing program. As data accumulated, problems became increasingly apparent. Air pollution, waste disposal, water supply, and mining methods emerged as serious roadblocks. On the Utah tracts, for example, the shale deposits were fractured in a manner that would prevent underground mining. On C-a optimal resource recovery could only be achieved by off-site disposal of spent shale (House Interior Committee, November 30, 1976, pp. 36–50). On one of the sites the air was so clean in one respect that no significant deterioration was allowed; yet for another pollutant background levels already exceeded ambient standards (Merrow, 1978).

During these two years of planning and study, the economics of shale development began to look less and less attractive. Operational problems and delay, aggravated by record-high inflation, forced repeated revision of cost estimates. The Utah group raised its estimate while it was still on the drawing boards from $600 million to $1.5 billion. At that rate, shale oil would only be profitable at $20 a barrel. Government price controls, meanwhile, were holding down domestic oil prices to less than $6 a barrel (White River Shale Project, August 1976, p. 16). "In February 1974," recalled one oil company executive, "there was a sense of urgency in the air as a result of the oil embargo and a certain momentum had started to build for a national energy policy that would support development of synthetic fuels." But by November of 1976 "this feeling [had] since disappeared," and with it,"the incentive for industry"(House Interior Committee, November 30, 1976, p. 48).

By year's end the Prototype Leasing Program was a shambles. Three participants had dropped out of the C-b group, the Utah project was effectively abandoned, and even the C-a project now looked infeasible. The secretary of the interior moved to suspend lease payments due to regulatory problems beyond control. Congress convened hearings to investigate the "footdragging," but no culprit, save the market, could be found (House Interior Committee, November 30, 1976, p. 57).

ENERGY SECURITY AND THE SECOND SHOCK

It was the Ayatollah, not Congress, that made the difference. In the spring of 1979 the Iranian Revolution not only triggered a second oil price shock but convinced most policy makers that dependence on foreign oil posed a serious threat to national security. This repetition of the first oil shock evoked widespread chagrin of "fool me once, shame on you; fool me twice, shame on me."

For the Carter administration, this second oil shock had two important implications. The National Energy Act, with its focus on conservation, had just passed Congress a few months earlier (Executive Office of the President, 1977, p. 60). Among other things, the Act had failed to resolve issues of domestic oil price controls. The onset of another crisis and the accident at Three Mile Island one month earlier convinced President Carter and his advisors that domestic oil prices should be decontrolled, and that alternative sources of domestic fuel should be vigorously encouraged. Here was a political "window of opportunity" for both (Schlesinger, March 1982).

In an address to the nation on April 5, 1979, President Carter announced that "now we must join together in a great national effort to use American technology to give us energy security in the years ahead" (Carter, 1980). First, domestic oil prices would be decontrolled gradually over eighteen months. To prevent domestic oil producers from capturing immense and undeserved profits, a special "windfall profits tax" would accompany price decontrol, raising an estimated $140 billion for the government over the next twelve years. The president also proposed an Energy Security Corporation that would use $88 billion of these revenues to subsidize development of 2.5 million barrels per day of synthetic fuels production.

The Energy Security Corporation was designed to avoid the government's past problems with pilot-plant contracting. First, it was conceived as a semi-autonomous, government-owned corporation on the model of the Reconstruction Finance Corporation. A board of directors and professional managers with a "business perspective" would use flexible, market-oriented incentives "to choose the most economic and expeditious approach" to synfuels. Loan and price guarantees, direct loans, and product purchases would be the preferred tools, although government-owned plants were not precluded. To short-circuit regulatory barriers and red tape, the president also proposed an Energy Mobilization Board to accompany the corporation (White House, August 1979).

President Carter's initiative and the ensuing escalation of oil prices from $14 to $32 a barrel induced a frenzy of legislative proposals and widespread debate over the wisdom of a crash program. Business interests agreed that something was needed, but felt that Carter's plan was too ambitious. Economists criticized the plan as "wasteful to the utmost, detrimental to the economic health of the country," and "unlikely to contribute significantly to solving our energy problems." Management experts objected on grounds of organizational limitations, bottlenecks, and shortages of manpower and infrastructure (Senate Banking Committee, July 1979, pp. 151–60; Joskow and Pindyck, 1979, pp. 18–24).

After a year-long battle over how the government's "windfall" should be divided up, Congress finally passed the Energy Security Act of 1980. The law created a U.S. Synthetic Fuels Corporation with a first-phase budgetary authority of $17 billion. After four years the corporation would submit a "comprehensive strategy" for congressional approval; only then would the balance of $68 billion become available (Public Law 96–294, n.d.; House Report, May 1980).

This massive commitment, together with the radical upturn in energy prices, convinced many of the largest oil and natural gas companies that the time for synfuels had come. With crude oil pushing $34 a barrel, oil shale looked especially feasible. Gulf and Standard of Ohio resumed work on their tract C-a. Exxon bought out Atlantic Richfield's stake in the Colony project for $400 million and announced its intention to build a $3 billion plant. And Union Oil of California announced a commercial plant supported by fast-track price guarantees from the Department of Energy. In North Dakota a consortium of gas companies began construction of the Great Plains Gasification Project with loan guarantees under the Defense Production Act (*Wall Street Journal*, June 18, 1980; "Synthetic Fuels," Summer 1980, pp. 3–11). But in March 1981, even before the Synfuels Corporation became operational or received senate approval of its board members, the price of Saudi light crude oil peaked at $36 and then slipped back to $32.

GLUT AND DISMANTLEMENT

This fourth programmatic attempt by the U.S. government to promote synthetic fuels failed more miserably than any of its predecessors. The Reagan administration rejected it from the outset and sought to eliminate its budget. Congress wrangled interminably over presidential appointments, over the salaries of the corporation's executives, and over its early practices and programs. But above all, the energy market had again turned to glut. In April 1986, as world oil prices plummeted below $12 a barrel, this prolonged political agony came to an end; three months earlier, Congress had voted to terminate the Synfuels Corporation. Only three projects received funding, and one of these, Great Plains Gasification, declared bankruptcy.

Forty years of historical experience with synthetic fuels policy testified to the extraordinary political impact of market forces. The cycle of surplus and glut in energy markets has clearly been the underlying factor in the formulation and implementation of governmental synfuels policy in America. The political process, moreover, seems to fluctuate according to market conditions between the two general models of political economy: a public-interest model during times of shortage where market failure appeared legitimate and a private-interest model during times of glut where interfuel benefits were at issue.

The explanation for these broad patterns lies in the institutional structure of the American economy, particularly the energy sector, and in the institutional structure of the political process, particularly business-government relations. Most markets in the United States, relative to other industrial countries, are dominated by large numbers of private firms. This is certainly the case in energy, where most other countries rely on state enterprise or at least a concentrated oligopoly in petroleum and electricity. Thus, in the United Kingdom, in France, and even in Japan, energy markets are not only less dynamic, but more easily controlled (Vietor, 1986). In the United States, however, the energy sector is composed of 20 vertically integrated, multinational oil companies, 100 refiners, several thousand independent producers, 25 gas pipeline companies, and more than 100 multibillion-dollar electric and gas utilities. In terms of supply and price, this sort of structure is difficult to control.

The reverse is true of the political structure. Administrative authority in the United States is generally less centralized than elsewhere, while legislative power is greater. Nowhere else has energy policy been so fractured (between 3 cabinet departments [Interior, Energy, and State], 2 federal regulatory agencies, and 50 state regulatory commissions), or have legislative committees taken such an active role in oversight.

As a result of these institutional circumstances, industry and the political system adjusted quickly to changes in the energy markets. And in view of the long lead times that development of synfuels entails, and the absence of a compelling national-security threat, none of the public programs worked. This record suggests two rather contradictory conclusions. Either government should stay out of the synfuels business altogether or else develop some new, more durable organizational format in which business and government can cooperate fully to modify the market. Hedging, it would seem, will not work.

NOTES

1. Because of Standard Oil's patent pool with I. G. Farben, the U.S. Alien Property Custodian had seized the German hydrogenation patents and made them available to the Bureau of Mines. Solicitor to M. Straus, September 9, 1944; H. H. Sargent to Straus, October 10, 1944; and M. Straus to H. L. Ickes, July 7, 1945; all in National Archives, Record Group 48, Department of the Interior, Central Classified Files, 1937–1953, box 3762, file 11–34. See also U.S. Congress, Senate Interior and Insular

Affairs Committee, January 1948, *Hearings on Synthetic Liquid Fuels: S134*, 80th Cong., 2nd sess., p. 60.

2. Brown's committee, whose members were appointed by the National Petroleum Council, would subsequently be the only security agency to oppose large-scale development of synthetic fuels. Bruce K. Brown to W. S. Hallanan, May 8, 1950, National Petroleum Council Files, file 017, "NPC Committee on Synfuel Liquefied Fuels Production Costs," 1950, Washington, D.C.: Department of the Interior. See also *National Petroleum News* 40 (February 4, 1948):20B; 40 (February 25, 1948):21.

3. Joseph Fisher, president of Resources for the Future, was appointed chairman of the Oil Shale Advisory Board. Other members were John Galbraith, the Harvard economist; Orlo Childs, dean of the Colorado School of Mines; H. Byron Mock, an attorney previously with the Bureau of Land Management; Benjamin Cohen, an attorney; and Milo Perkins, a financial consultant. Mr. Cohen and Mr. Perkins had served in government during the New Deal.

4. J. R. Freeman, editor of a local newspaper in Colorado, received considerable attention and the 1967 Herric Editorial Award for his series of 40 articles on the "Grabber Conspiracy" to exploit oil shale. See also C. Moore to N. Peterson, January 12, 1967, and draft to J. R. Freeman, in Cordell Moore Papers, Chronological File, box 3, Austin, Tex.: Lyndon B. Johnson Presidential Library.

5. The National Environmental Policy Act, January 1, 1970, required that environmental impact statements be prepared for all federally funded programs that impact the environment. See Natural Resources Defense Council, February 29, 1973, *Comments on Prototype Oil Shale Leasing Program, Draft Environmental Impact Statement*, Oil Shale Files, Oil Shale, pt. 7, Washington, D.C.: U.S. Department of the Interior, p. 25. See also The Institute of Ecology, October 29, 1973, *Environmental Impact Assessment Project: A Scientific and Policy Review of the Prototype Oil Shale Leasing Program Final Environmental Impact Statement*, Washington, D.C.: Environmental Impact Assessment Project.

REFERENCES

Agreement between the United States Department of the Interior and the American Gas Association for the Cooperative Coal Gasification Research Program, August 3, 1971. June 1973. In U.S. Congress, Senate, Interior Committee, Energy Research and Development Policy Act. 93d Cong., 1st sess., pp. 652–58.

American Gas Association. June 1973. *Coal Gasification Pilot Plant Research Program, Third Quarter Fiscal Year 1973*. In U.S. Congress, Senate, Interior Committee, Energy Research and Development Policy Act. 93d Cong., 1st sess., pp. 652–58.

Anderson, Douglas D. 1981. *Regulatory Politics and Electric Utilities*. Boston: Auburn House.

Baumol, William. 1965. *Welfare Economics and the Theory of the State*. Cambridge, Mass.: Harvard University Press.

Boyd, James. January 1948. *Summary of Appropriations Requests*. In U.S. Congress, Senate, Interior Committee, Hearings on Synthetic Liquid Fuels: S. 134. 80th Cong., 2d sess., pp. 25–65.

Brief History of Cresap, West Virginia Pilot Plant Operations. November 9, 1971. Office of Coal Research Contract Files, box 11, file no. 5. Suitland, Md.: Federal Records Center.

Brown, B. June 27, 1950. Letter to W.S.S. Rodgers. Department of the Interior, National Petroleum Council Files, file 017, 1950.

Carter, Jimmy. 1980. Energy Address to the Nation, April 5, 1979. *Public Papers of President Jimmy Carter. Vol. 1.* Washington, D.C.: Government Printing Office.

Chapman, O. April 21, 1950. Letter to W. S. Hallanan. National Petroleum Council Files, file 106.61, 1951. Washington, D.C.: Department of the Interior.

Charmbury, H. R. December 13, 1961. Letter to G. Lamb. General Technical Advisory Committee Correspondence Files, 1960–1962, box 1, Office of Coal Research Accession no. 48–74–12. Suitland, Md.: Federal Records Center.

Chief, Division of Contracts and Administration. November 27, 1967. Memorandum to C. Moore. Office of Coal Research Procurement Files, pt. 10. Washington, D.C.: Department of the Interior.

Congressional Record. 1953. Vol. 99, pt. 3.

Cooper, J. S. May 21, 1954. Letter to D. D. Eisenhower. Eisenhower Library, White House Office File, box 678, file 134-E, undated.

Davies, R. K. August 1943. In U.S. Congress, Senate, Interior Committee, Subcommittee on Public Lands. *Hearings on Synthetic Liquid Fuels*. 78th Cong., 1st sess.

Day, R. February 24, 1949. Letter to J. A. Krug. National Archives, Record Group 48, Department of the Interior, Central Classified Files, 1937–1953, box 3763.

Dole, H. May 21, 1969. Letter to R. Byrd. Office of Coal Research, Equipment and Property Files, pt. 1. Suitland, Md.: Federal Records Center.

Douglas, P. August 6, 1968. Letter to S. Udall. Oil Shale Files, pt. 16. Washington, D.C.: Department of the Interior.

Eisenhower, Dwight D. September 16, 1959. Memorandum of Disapproval of Bill Creating a Coal Research and Development Commission. *Public Papers of the Presidents, Dwight David Eisenhower*. Washington, D.C.: Government Printing Office, p. 660.

Elrond, F. S. December 19, 1961. Letter to G. Lamb. General Technical Advisory Committee Correspondence Files, 1960–1962, box 1, Office of Coal Research, Accession no. 48–74–12. Suitland, Md.: Federal Records Center.

Executive Office of the President. 1977. *The National Energy Plan.*Washington, D.C.: Government Printing Office.

Friedman, R. February 23, 1949. Letter to R. Day. National Archives, Record Group 48, Department of the Interior, Central Classified Files, 1937–1953, box 3763.

Fumich, G. December 11, 1967. Letter to Secretary of the Interior. Office of Coal Research Procurement Files, pt. 10. Washington, D.C.: Department of the Interior.

General Technical Advisory Committee. June 6, 1961. GTAC Meeting Minutes. GTAC Correspondence Files, 1960–1962, box 1, Office of Coal Research Accession no. 48–74–12. Suitland, Md.: Federal Records Center.

———. January 27, 1965a. GTAC Meeting Transcripts. GTAC Correspondence Files, 1965, box 1, Office of Coal Research Accession no. 84–74–12. Suitland, Md.: Federal Records Center.

———. January 27, 1965b. GTAC Meeting Transcripts. GTAC Correspondence

Files, 1965, box 2, Office of Coal Research Accession no. 48–74–12. Suitland, Md.: Federal Records Center.

———. July 13, 1965. GTAC Meeting Transcripts. GTAC Correspondence Files, 1960–1962, box 1, Office of Coal Research Accession no. 48–74–12. Suitland, Md.: Federal Records Center.

———. October 3, 1966. GTAC Meeting Transcripts. GTAC Correspondence Files, 1966, box 2, Office of Coal Research Accession no. 48–74–12. Suitland, Md.: Federal Records Center.

———. May 21, 1968. GTAC Meeting Transcript. GTAC Correspondence Files, 1968, box 2, Office of Coal Research Accession no. 48–74–12. Suitland, Md.: Federal Records Center.

———. October 7, 1969. GTAC Meeting Transcript. GTAC Correspondence Files, 1969, box 3, Office of Coal Research Accession no. 48–74–12. Suitland, Md.: Federal Records Center.

———. January 12, 1971. GTAC Meeting Transcript. GTAC Correspondence Files, 1971, box 4, Office of Coal Research Accession no. 48–74–12. Suitland, Md.: Federal Records Center.

Gulf Research and Development Co. December 1982. Interviews with Richard H. K. Vietor. Pittsburgh.

Hardy, R. A. March 4, 1959. Letter to W. N. Aspinall. U.S. Congress, House. Report no. 370. 86th Cong., 1st sess.

———. December 29, 1959. Letter to Secretary of the Interior. White House Office File, box 678, file 134-H. Abilene, Kans.: Dwight D. Eisenhower Presidential Library.

Hughes, Thomas. August 1969. "Technological Momentum: Hydrogenation in Germany, 1900–1933." *Past and Present* 55:106–32.

Ickes, H. L. January 29, 1943. Letter to M. Straus. National Archives, Record Group 48, Department of the Interior, Central Classified Files, 1937–1953, box 3742.

Ickes v. Virginia-Colorado Development Co., 1935. 295 U.S. 639.

Joskow, Paul, and Pindyck, Robert. 1979. "Synthetic Fuels." *Regulation*, pp. 18–42.

Kelley, John M. April 30, 1964. Remarks by Assistant Secretary of the Interior John M. Kelly before the First Annual Oil Shale Symposium, Colorado School of Mines, Golden, Colorado. Oil and Gas Division File 403.3. Washington, D.C.: Department of the Interior.

Kerr, K. Austin. 1968. *American Railroad Politics, 1914–1920*. Pittsburgh: University of Pittsburgh Press.

Kolko, Gabriel. 1963. *The Triumph of Conservatism: A Reinterpretation of American History, 1900–1916*. New York: Glencoe Press.

Krammer, Arnold. July 1978. "Fueling the Third Reich." *Technology and Culture* 19: 394–422.

Krug, J. A. February 9, 1949. Letter to K. Wallgren. National Archives, Record Group 48, Department of the Interior, Central Classified Files, 1937–1953, box 3763.

Larson, G. F. July 2, 1963. Letter to Director of Budget. Office of Coal Research Fiscal Year 1964 Budget File, box 18, Department of the Interior Accession no. 71A–963. Suitland, Md.: Federal Records Center.

Lehner, S. December 13, 1961. Letter to G. Lamb. General Technical Advisory

Committee Correspondence Files, 1960–1962, box 1, Office of Coal Research Accession no. 48–74–12. Suitland, Md.: Federal Records Center.

Merrow, Edward W. 1978. *Constraints on the Commercialization of Oil Shale*. Santa Monica, Calif.: Rand Corporation.

Minutes of Meeting with Senator Robert Byrd, March 21, 1969. Consolidation Coal, Project Gasoline File, box 15, Office of Coal Research Accession no. 48–73–11. Suitland, Md.: Federal Records Center.

Monthly Operations Log. Undated. Office of Coal Research, Equipment and Property Files, pt. 5. Washington, D.C.: Department of the Interior.

Moore, J. C. December 2, 1968. Remarks to the Oklahoma Independent Petroleum Association, Tulsa, Oklahoma. Cordell Moore Papers, box 1: Oil Imports and National Security. Austin, Tex.: Lyndon B. Johnson Presidential Library.

Morton, R. January 10, 1972. Letter to H. R. Sharbaugh, President, Sun Oil Co. Oil Shale Files, pt. 7: Oil Shale. Washington, D.C.: Department of the Interior.

Nash, Gerald. 1968. *United States Oil Policy, 1890–1964*. Pittsburgh: University of Pittsburgh Press.

National Petroleum Council. 1951. *Report of the National Petroleum Council's Committee on Synthetic Liquid Fields Production Costs*. Synthetic Fuels File. Washington, D.C.: Department of the Interior.

———. January 29, 1952. *Interim Report of the National Petroleum Council's Committee on Synthetic Fuels Production Costs*. Washington, D.C.: Department of the Interior.

———. July 25, 1952. *Interim Report of the National Petroleum Committee on Liquid Fuels Production Costs*. p. 8.

———. July 29, 1952. Meeting Transcript. Synthetic Fuels File, Washington, D.C.: Department of the Interior.

———. February 26, 1953. *Final Report on Synthetic Liquid Fuels Costs*. National Petroleum Council File. Washington, D.C.: Department of the Interior.

———. January 25, 1955. *Report of the National Petroleum Council's Committee on Oil Shale Policy*. National Petroleum Council Files, file 106.61, 1955. Washington, D.C.: Department of the Interior.

———. May 5, 1955. Meeting Transcript. Synthetic Fuels File. Washington, D.C.: Department of the Interior.

National Petroleum News. January 21, 1948.

New York Times. September 12, 1948.

Nixon, Richard M. June 4, 1971. Presidential Energy Message to Congress. *Energy Management*. Washington, D.C.: Commerce Clearing House, 1973.

Office of Coal Research. February 18, 1962. *Contract Information Projections*. Office of Coal Research Fiscal Year 1963 Budget File, box 17, Department of the Interior Accession no. 71A–963. Suitland, Md.: Federal Documents Center.

———. 1968. *Annual Report 1967*. Washington, D.C.: Department of the Interior.

———. 1969. *Annual Report 1968*. Washington,D.C.: Department of the Interior.

———. 1973. *Annual Report 1972*. Washington, D.C.: Department of the Interior.

Oil Shale Advisory Board. February 1, 1965. *Interim Report*. Washington, D.C.: Department of the Interior.

"Oil Shale and Shale Oil: A Survey, Part II." April 29, 1943. *Oil and Gas Journal*, pp. 73–79.

Olson, Mancur. 1965. *The Logic of Collective Action: Public Action and the Theory of Groups*. Cambridge, Mass.: Harvard University Press.
Ottinger, Richard. 1976. In U.S. Congress, House, Committee on Interstate and Foreign Commerce, Subcommittee on Energy and Power. *Synthetic Fuel Loan Guarantees*. 94th Cong., 2d sess., vol. 1, pp. 77–82.
Patterson, R. July 2, 1947. Telephone Memo to J. Krug. National Archives Record Group 48, Department of the Interior, Central Classified Files, 1937–1953, box 3763.
Porter, Michael, and Sagansky, Jeffrey. 1976. "Information, Politics, and Economic Analysis: The Regulatory Decision Process in the Air Freight Cases." *Public Policy* 24:263–307.
Proxmire, William. August 1968. Letter to S. Udall. Oil Shale Files, pt. 16. Washington, D.C.: Department of the Interior.
Public Law 93–438. October 11, 1974. U.S. Congress, House, Committee on Science and Technology. *Principal Energy Research and Development Legislation*. 95th Cong., 2d sess.
Public Law 96–24. 1980. 96th Cong., 2d sess.
Reichl, Eric. December 1980. Interview with Richard H. K. Vietor.
Report of the Interdepartmental Committee on the Soft Coal Industry. October 7, 1960. Fred Seaton Papers, box 2, Minerals/Coal Program. Abilene, Kans.: Dwight D. Eisenhower Presidential Library.
Review of the Coal Research Program. Undated. Office of Coal Research Budgetary Material. Suitland, Md.: Federal Records Center.
Schlesinger, James. March 1982. Interview with Richard H. K. Vietor.
Shell Oil, Phillips Petroleum, Atlantic Refining, Union Oil, and TOSCO. January–April, 1964. Correspondence with Stewart Udall. Oil Shale Files, pts. 1–2. Washington, D.C.: Department of the Interior.
Sporn, P. December 21, 1961. Letter to G. Lamb. General Technical Advisory Committee Correspondence Files, 1960–1962, box 1, Office of Coal Research Accession no. 48–74–12. Suitland, Md.: Federal Records Center.
Steele, Henry. 1967. The Prospects for the Development of a Shale Oil Industry. In U.S. Congress, Senate, Committee on the Judiciary, Subcommittee on Antitrust and Monopoly. *Competitive Aspects of Oil Shale Development*. 90th Cong., 1st sess., pp. 542–58.
Stigler, George J. 1971. "The Theory of Economic Regulation." *Bell Journal of Economics and Management Science* 2:3–21.
Straus, M. January 28, 1943. Letter to H. L. Ickes. National Archives, Record Group 48, Department of the Interior. Central Classified Files, 1937–1953, box 3762, file 11–34.
———. February 1, 1943. Letter to R. K. Davies. National Archives, Record Group 48, Department of the Interior, Central Classified Files, 1937–1953, box 3762.
———. April 3 and August 20, 1943. Letters to H. L. Ickes. National Archives, Record Group 48, Department of the Interior, Central Classified Files, 1937–1953, box 3762.
———. January 27, 1944. Letter to J. Guffey. National Archives, Record Group 48, Department of the Interior, Central Classified Files, 1937–1953, box 3762.
"Synthetic Fuels: The Processes, Problems and Potential." Summer 1980. *The Lamp*, pp. 3–11.

Udall, Stewart. 1963. *The Quiet Crisis*. New York: Avon Books.

———. January 18, 1968. Letter to John Dingell. Oil Shale Files, pt. 15. Washington, D.C.: Department of the Interior.

———. October 24, 1968. Letter to S. Hathaway. Oil Shale Files, pt. 17. Washington, D.C.: Department of the Interior.

Union Oil Company of California. 1964. 71 I.D. 169.

U.S. Bureau of Mines. 1949a. *Report of Investigations, No. 4564: Estimated Plant and Operating Cost for Producing Gasoline from Coal Hydrogenation*. Washington, D.C.: Government Printing Office.

———. 1949b. *Synthetic Liquid Fuels: Annual Report of the Secretary of the Interior*. Washington, D.C.: Government Printing Office.

———. 1950. *Annual Report of the Secretary of the Interior*. Washington, D.C.: Government Printing Office.

———. 1951. *Annual Report of the Secretary of the Interior*. Washington, D.C.: Government Printing Office.

———. October 25, 1951a. *Comments on Reports of the National Petroleum Council Subcommittee on Synthetic Liquid Fuels Production Costs for Oil Shale*. Washington, D.C.: Government Printing Office.

———. October 25, 1951b. *Cost Estimate for Coal Hydrogenation*. Washington, D.C.: Government Printing Office.

U.S. Congress. House. 1956. House Resolution 400. 84th Cong., 2d sess.

———. House. May 1980. *Report No. 96-1104: Conference Report on Energy Security Act*. 96th Cong., 2d sess.

———. House Committee on Appropriations. March 11, 1953. *Hearings on Interior Department Appropriations for 1954*. 83d Cong., 1st sess., pt. 2.

———. House Committee on Interior and Insular Affairs. February 4, 1960. *Report No. 1241*. 86th Cong., 2d sess.

———. House Committee on Interior and Insular Affairs. 1961. *Politics, Programs, and Activities of the Department of the Interior*. 87th Cong., 1st sess.

———. House Committee on Interior and Insular Affairs. November 30, 1976. *Oversight—Prototype Oil Shale Leasing*. 94th Cong., 2d sess.

———. House Committee on Interior and Insular Affairs, Special Subcommittee on Coal Research. 1957a. *Report No. 1263*. 85th Cong., 1st sess.

———. House Committee on Interior and Insular Affairs, Special Subcommittee on Coal Research. 1957b. *Coal*. 85th Cong., 1st sess., pt. 2.

———. House Committee on Interstate and Foreign Commerce. March 1948. *H.R. 5475: Committee Print*. 80th Cong., 2d sess.

———. House Committee on Science and Astronautics, Subcommittee on Science, Research, and Development. April 25, 1969. *House Document No. 91-137: Technical Information for Congress*. 91st Cong., 1st sess.

———. House Committee on Science and Technology. December 1978. *Principal Energy Research and Development Legislation*. 95th Cong., 2nd sess.

———. Senate Committee on Appropriations, Subcommittee on Department of the Interior Appropriations. February 24, 1964. *Interior Department and Related Agencies Appropriations for 1965*. 88th Cong., 2d sess.

———. Senate Committee on Banking, Housing, and Urban Affairs. July 1979. *Energy Financing Legislation*. 96th Cong., 1st sess.

———. Senate Committee on Interior and Insular Affairs. January 1948. *Hearings on Synthetic Liquid Fuels: S. 134*. 80th Cong., 2d sess.
———. Senate Committee on Interior and Insular Affairs. 1973. *Energy Research and Development Policy Act*. 93d Cong., 1st sess.
———. Senate Committee on Interior and Insular Affairs, Subcommittee on Minerals, Materials, and Fuels. June 1959. *Coal Research*. 86th Cong., 1st sess.
———. Senate Committee on Interior and Insular Affairs, Subcommittee on Minerals, Materials, and Fuels. November 15, 1971. *Prototype Oil Shale Leasing Program, Hearings: Oil Shale*. 92d Cong., 1st sess.
———. Senate Committee on Interior and Insular Affairs, Subcommittee on Minerals, Materials, and Fuels. March 1976. *Hearings: Oil Shale Leasing*. 94th Cong., 2d sess.
———. Senate Committee on the Judiciary, Subcommittee on Antitrust and Monopoly. April–May 1967. *Competitive Aspects of Oil Shale Development*. 90th Cong., 1st sess.
———. Senate Special Committee Investigating Petroleum Resources. January 31, 1947. *Report No. 9: Investigation of Petroleum Resources in Relation to the National Welfare*. 80th Cong., 1st sess.
U.S. Department of Energy. August 1978. *Fossil Energy Program Report*. DOE/ET0060–78. Washington, D.C.: Government Printing Office.
U.S. Department of Interior. 1955. National Petroleum Council File.
———. July 7, 1964. *The Oil Shale Policy Problem*. Oil Shale Files, pt. 4. Washington, D.C.: Department of the Interior.
———. January 27, 1967. Press Release, Office of Secretary. Synthetic Liquid Fuels (1966–67), Oil Shale Files, file 403.3. Washington, D.C.: Department of the Interior.
———. 1968. *Prospects for Oil Shale Development*. Washington, D.C.: Department of the Interior.
———. December 27, 1968. Press Release, Office of Secretary. Oil Shale Files, pt. 18. Washington, D.C.: Department of the Interior.
———. January 1969–July 1970. Correspondence Files, Office of the Secretary, Central Classified Files, Minerals and Fuels/Commodities and Products, Oil Shale Files, pts. 1 and 2: Oil Shale. Washington, D.C.: Department of the Interior.
———. April 21, 1969. Memorandum for the Record, Meeting between Officials of Department of the Interior, Continental Oil, and Consolidation Coal. Office of Coal Research, Equipment and Property Files, pt. 1. Washington, D.C.: Department of the Interior.
———. November 28, 1973. *Decision Statement of the Secretary of the Interior on the Prototype Oil Shale Leasing Program*. In U.S. Congress, Senate, Committee on Interior and Insular Affairs, Subcommittee on Minerals, Materials and Fuels. 1973. *Prototype Oil Shale Leasing Program*. 93d Cong., 1st sess.
U.S. Office of Management and Budget. December 18, 1970. *Circular A-100*. Office of Coal Research, Equipment and Property Files, pt. 3. Washington, D.C.: Department of the Interior.
Vietor, Richard H. K. 1984. *Energy Policy in America since 1945*. New York: Cambridge University Press.
———. 1986. "Energy Markets and Policy." In *America Versus Japan: A Comparative*

Study, ed. by Thomas McCraw. Boston: HBS Press, pp. 44–63, 163–189, 324–340.

Wall Street Journal. June 18, 1980.

Welles, Chris. 1970. *The Elusive Bonanza*. New York: E. P. Dutton.

White House. August 1979. *The President's Program for United States Energy Security: The Energy Security Corporation*. Washington, D.C.: Office of the Presidency.

"White River Shale Project: A Position of Oil-Shale Development." August 1976. *Shale Country* 2:16.

Wilson, R. July 12, 1944. *The Technical and Economic Status of Liquid Fuel Production from Non-Petroleum Sources, with Recommendations as to a Research Program*. File labelled: Post-War Army and Navy, July 1–12, 1944. Cambridge, Mass.: Massachusetts Institute of Technology Archives, Compton-Killean Papers.

2

Synthetic Fuel Technology Nondevelopment and the Hiatus Effect: The Implications of Inconsistent Public Policy

MICHAEL M. CROW

> As oil price rises, projected costs of producing synfuel from a new planned plant using currently foreseeable technology increase proportionately. No matter how high the price of oil rises—even to $100 per barrel—a new plant built subsequent to arrival of oil at that price will not be economic as an investment prospect.
>
> —Congressional Research Service, 1981, p. 1

With the demise of the Synthetic Fuels Corporation in 1986,[1] the 90 percent reduction of federal synthetic-fuel research funding from 1981 onwards,[2] and the post–1983[3] shutdown of the major synthetic fuels research laboratories in industry, the United States ended its fourth major synthetic fuels technology (SFT) development thrust in this century without having demonstrated the commercial viability of a single process configuration.[4] To many observers, the underlying reason for this technology-development failure is pure and simple economics. If synthetic fuels cannot be produced for less than the cost of depletable natural fuels, then money should not be wasted in the effort.

There is nothing new or original in this explanation. As early as the 1950s the National Petroleum Council (NPC) concluded that synthetic liquids from coal could not economically compete with petroleum. In attacking the cost estimates of the U.S. Bureau of Mines R&D program, NPC concluded:

> The need for a synthetic liquid fuel industry in this country is still in the distant future. Since new techniques may be available then, we question the wisdom of the Government financing large-scale demonstration plants. (NPC, 1953)

In contrast to the economics issue, Vietor (1984, 1980) points to poor public-sector/private-sector interaction in the SFT development process experience as a major cause of the failure. Others (Desai and Crow, 1983; Crow and Hager, 1985) have concluded that the politics surrounding the decision making and plant siting, as well as the "crisis" environment mobilized, were the key contributing variables.

Unfortunately, none of these explanations addresses the issue of why the technology itself has failed to develop much beyond the first generation efforts of the 1920s and why the technology has continuously been plagued by failure and cost overruns (GAO, 1977; Congressional Research Service, 1980a and 1980b; Vietor, 1980; Landsberg and Coda, 1983). Only a few authors have addressed the more fundamental issue: Despite the four major national policy initiatives to stimulate a synthetic fuels industry in the United States and the several billion dollars spent on this technology since 1940, there has been little real progress in developing the synthetic fuels technology or in understanding the basic science of the processes and feedstocks associated with the technology (National Research Council, 1980; Whitehurst, 1980; Lee, 1982; IESCES Symposium, 1984). Process efficiencies have slightly improved, but there have been no major breakthroughs in the technology itself since its earliest days (Polaert, 1985; Science News, 1983; Cooper, 1981). The central hypothesis of this chapter is that the lack of continuous technical progress associated with the technology has been the principal barrier to development, not the economic, interorganizational, or political problems associated with SFT development.

None of the traditional explanations for SFT failure seems to recognize that each of the major problems mentioned above has been overcome in similar macro-engineering technology development efforts.[5] Considering major technology efforts of the last 45 years, it is fair to say that most of these projects experienced extensive political decision-making difficulties (SST), problems with the market economy (space shuttle, nuclear power, and SST), and problems related to the public-sector/private-sector interface (nuclear power and Apollo). While the characteristics of the problems encountered differ among individual technologies, the nature of the problem encountered by developers has been, and will probably continue to be, similar.

Horwitch (1979), Seamans and Ordway (1977), and Sayles and Chandler (1971) indicate that the typical problems associated with the pursuit of any new macro-engineering project include

1. coordination between the public and private sector
2. complex policy environments
3. uncertain economics
4. interorganizational managerial problems
5. need for adequate steady funding
6. decentralized decision making

These problems appear to be the result of the complex nature of the technologies being developed and the fact that the technical and organizational scope of these endeavors requires significant interorganizational coordination between the private and public sectors. On the basis of this literature, I think it fair to assume that the development failure associated with synthetic fuels technology in the United States is at least partially explained by the economics of oil, the problems of sectoral coordination, and the politics of the energy crisis.[6] If an improved understanding of large-scale technology development processes is to be developed and if the potential of SFT is to be more accurately assessed, it is necessary to explain why these problems became more pervasive in the case of SFT development than in the case of other macro-engineering technologies.

To address this gap in the literature, this chapter will attempt to demonstrate that, in addition to the familiar problems associated with synthetic fuels development, this technology suffered from a major problem of discontinuous scientific and engineering activity. In effect, SFT technology in the United States has experienced several major hiatuses where little or no significant research, development, or demonstration activity was carried out or where insufficient time and effort were devoted to the technology. These hiatus periods, which were the product of inconsistent public policy[7] and underinvestment in basic research by industry and government, had the result of limiting the potential of SFT as a fuel replacement option. Given the fact that the United States has just entered another major technology-development hiatus, it is important that the potential effect of previous periods on subsequent synthetic fuels development be understood prior to the initiation of a future research development and demonstration (RD&D) program.

EXPLANATIONS FOR SYNTHETIC-FUEL TECHNOLOGY FAILURE

To understand better the difficulties associated with national efforts to develop SFT, a review of the competing explanations for the failure of the technology-development process is required. The various explanations are well known and have often been discussed individually as the principal reason for the difficulties and failures associated with demonstrating SFT. These explanations include the following:

The Price of Oil: The Moving Target

Petroleum is a commodity that has unique supply and demand characteristics. These characteristics, when combined with uncertainties of the non–United States petroleum-producing community, play an important role in increasing the levels of uncertainty related to the development of any replacement fuels technology. These uncertainties, particularly the price uncertainties and the financial-scale imponderables of a synthetic-fuel industry, are so large that standard market-based investment is impossible (Rothberg, 1984). The financial risks are simply too great given the fluctuations in oil price. Furthermore, the Congressional Research Service (1981) has argued that the price of SFT production is at least partly a function of the price of oil.

A key implication of this is that the economics of petroleum production and distribution prohibit the development of synthetic fuels without large-scale government intervention. Given the fact that government intervention in the synthetic-fuel development cycle has only occurred in response to supply or price fluctuations or in periods of increased central planning[8] and given that government efforts have always been terminated with the stabilization of the petroleum economy, one may rightly conclude that the economics of petroleum is a critical variable affecting synthetic-fuel technology development. Despite its explanatory force, the price of oil does not alone account for the failure of SFT in the United States, particularly given the RD&D funding and market support efforts of the government.

While certainly important from a financial investment perspective, the economics variable is perhaps more controlling in the way it forces an underinvestment in R&D related to synthetic fuels. In his classic article on the economics of basic research, Nelson (1959) demonstrates that industries left to their own devices in a market economy have little incentive to carry out basic research. The result is that without government funding of basic research to underpin development of a technology, technological R&D will not occur on a scale sufficient to address the limiting problems of the developing technology. Thus, the economics explanation is a likely contributing factor to the knowledge-production rate. This point is particularly important to SFT development.

Because of the nature of the economics of petroleum and the unpredictability of supply and price fluctuations, government intervention has focused on the goal of rapid commercialization and production. As a result, production of basic knowledge to support the development of SFT has never received consistent funding.[9] Therefore, the price of oil explanation also explains the discontinuous nature of the basic research associated with synthetic-fuel technology development.

The Struggle between the Collective Good and the Private Interest

Vietor's seminal text, *Energy Policy in America since 1945* (1984), reviews the development history of synthetic fuels technology since 1945 within the general energy policy context. With only minimal hesitation Vietor concludes that the failure to achieve SFT in the United States was the result of a "failure of institutional relations that left administrative agencies without the financial or technical resources, and especially the political autonomy, necessary to get any kind of job done" (1984, p. 346). This failure of "institutional relations" was the result of the ongoing conflict between what Goodwin (1981) saw as the somewhat schizophrenic character of U.S. energy policy and its unresolved free-market/central-planning conflict. This ongoing conflict, when blended with the pork-barrel character of many demonstration-project selections, acted as a deterrent to sectoral cooperation. Consequently, a unified program plan involving industry, the federal bureaucracy, and Congress was never developed. Because the means for producing such a program were never provided, the proper roles for government, industry, and the market were never defined.

The result of this coordination failure was that in 1981 the Reagan administration was left no choice but to cut the government's losses and to eliminate the entire SFT development effort. Faced with the choice between funding continued nondevelopment and letting the market run its own course, the Reagan administration chose the latter as a means of ending the government's tendency to ignore the market (Crow, 1982). Clearly, then, the lack of a consistent program plan between industry and government, compounded by the lack of coordination between industrial and governmental R&D personnel even when the industrial personnel were under contract,[10] contributed to the discontinuous nature of the development process.

Synthetic Fuels Technology Policy: Pork, Panic, and Politics

The shaping of public policy designed to stimulate the development of SFT has been fraught with problems. These have included (1) the tendency to underutilize scientific and technical information in SFT R&D decisions (Desai and Crow, 1983); (2) the use of demonstration plants as a means of political risk versus technical risk reduction (Crow and Hager, 1985); (3) the generation of multiple policy goals associated with SFT development (Vietor, 1980, 1985; Kash and Rycroft, 1984); (4) the inability of government to pick"winning" technologies (Nelson and Langlois, 1983); and (5) the general problem of goal setting in the SFT policy arena (Rudolph and Willis, 1985).

When considered collectively, these policy-making problems lead to the conclusion that government has often attempted to do too much without enough information in an effort to address different and often conflicting goals. It is simply impossible to set production goals for a SFT program that lacks a knowledge base beyond its first generation and lacks the operating experience from which realistic expectations can be developed. In addition, when the siting of the demonstration plant is a function of political geography as much as it is a function of engineering criteria, it is to be expected that there will be goal conflict problems and technical trade-offs that could jeopardize the technical success and ultimately the economics of a process.[11]

The problem of policy making is particularly acute as a result of what Vietor (1985) calls the the interfuel politics between oil and coal interests. Prior to the major acquisition of coal resources by the oil industry, the coal industry and the oil industry were very different types of businesses and were in substantial conflict with each other during many of the SFT policy-making periods. Furthermore, the annual reports of the Office of Coal Research (1968–1974) very clearly illustrated that there existed within the leadership of the coal industry a strong tendency to support only short-term R&D activities. The industry's feeling was that if the R&D project could not help the coal industry improve markets or lower costs in 2–3 years, then it was not worth pursuing.

As a result of the factors mentioned above and the drastic change toward a free-market development policy in 1980 (Crow, 1982; Kash and Rycroft, 1984), SFT development policy has been very disjointed, unclear, and often illogical from a scientific perspective. The result has been the premature funding of demonstration plants (Horwitch, 1980; Rothberg, 1984; Stanfield, 1984) and a general failure of SFT development efforts.

Premature Development: Too Much Development and Not Enough Research

In 1969 the United States demonstrated a first-generation interplanetary space technology with the Apollo 11 landing on the moon. To most observers this technology-demonstration project was initiated by President Kennedy in his 1962 race to the moon speech.[12] In actuality, the 1969 U.S. demonstration was the result of over 45 years of continuous basic research and 25 years of technology-development and demonstration efforts in a wide range of supporting fields (McDougall, 1985; Bilstein, 1980). From the early 1920s in the United States and even earlier in Europe, a continuously funded R&D program in rocketry and later in space flight had been financed by various governments (Roland, 1985; Stares, 1985). The fruit of these efforts was that even before President Kennedy's announcement of the goal of reaching the moon by 1970, the principal components of the technology were already well developed (Steinberg, 1985; Bilstein, 1980). Thus, the technical risks, while certainly present, were known and were seen as manageable.

By contrast, the United States has yet to demonstrate a commercially viable synthetic-fuel process. This failure, as discussed earlier, has not been the result of insufficient funding or inadequate time. For example, President Ford's initial commitment to SFT was made in 1975 with the Project Independence effort (ERDA, 1975). The lack of development success was a function of the inadequacy of predevelopmental research for SFT (Wang, 1981; International Energy Agency, 1982; Desai and Crow, 1983; Penner, 1981).

In his analysis of why certain large-scale technological projects succeed while others fail, Steinberg (1985) found that those macro-engineering projects that failed had certain characteristics, all of which seem applicable to the SFT case. First, most of these projects arose as a means of accelerating technology development toward a national objective. Second, due to the accelerated development goal, the relatively orderly process of step-by-step technology development was replaced by a parallel development activity involving simultaneous R&D on numerous aspects of the technology. Third, because of the parallel nature of the development projects, any failure to overcome a technological obstacle on one path would probably result in a delay or failure of the entire effort. Consequently, investments in the pursuit of other parallel tasks would likely be wasted.

A review of eight major technology-development projects (Apollo, Manhattan, Polaris, MX Basing, Safeguard Ballistic Missile Defense, Nuclear Airplane, Starwars, and Skybolt) revealed that those projects that initiated parallel development efforts before the basic technology was proven through research always failed and that those that had an established research base tended to be successful. When a second dimension of a static or dynamic objective is added, the probability of development failure increased with more dynamic objectives. Under Steinberg's definition SFT would be classified as a dynamic objective technology because of the "crisis" nature of the goals imputed to SFT development efforts.

Beyond the evidence presented earlier, the inadequacy of the basic knowledge core for the development of SFT can also be cited. The problems at the Catlettsburg H-Coal demonstration plant (Shangraw, December 1982), as well as those at other demonstrations (Vick and Epperly, 1982; O'Hara, 1979), illustrate the effect of placing too much emphasis on development and not enough on research. If we use the Steinberg model for differentiating successful from unsuccessful ventures, it is clear that an inadequate knowledge base made the failure of SFT predictable.

THE HIATUS EFFECT ON TECHNOLOGY DEVELOPMENT

Each of the explanations for the failure of SFT is, in and of itself, plausible and defensible. Each, however, only partly explains the problem faced by SFT. SFT processes, as conceptualized in the periods of U.S. development, were not competitive given the price of oil and the market economics of large oil/energy companies. The result was that each effort to develop SFT ended in failure. But does this explanation indicate why the technology itself failed? No.

Recent analyses by the Rand Corporation (Hess, 1985a and 1985b; Myers and Arguden, 1984; Merrow, Phillips, and Myers, 1981) indicate that a critical variable in the success of any process-oriented technology is the *continuous* support of a RD&D program over an extended period of time. Discontinuous support or a break in the process-development cycle, labeled the hiatus effect, was viewed as the most critical factor in determining the success probability of process-type technologies like SFT. Furthermore, an analysis of information/knowledge survival rates, both in private firms and in the government, suggests that information and knowledge of technologies, such as SFT, are often lost during periods of non–R&D activity. This loss factor appears to be a function of the extent to which new demonstration and pilot-plant data are produced on a continuous basis. Another important factor identified by the Rand team is the importance of person-to-person knowledge transfer in this type of R&D effort. Apparently, after an R&D team is dismantled and assigned new duties, the loss rate of knowledge is very high.

In addition to the effect of hiatus periods on the loss of information and knowledge important to the process-development cycle, the improvement of plant performance in process-oriented technology also appears to be a function of continuous incremental improvements (Hess, 1985a). Using the SASOL I experience, it is estimated that it takes 4–5 years to eliminate basic performance problems in innovative pioneer plants. This suggests, according to Hess, that "for an innovative technology, a second plant potentially faces fewer startup difficulties if its design *does not begin* until after the pioneer plant has had at least 5 years of operational experience" (Hess, 1985a, p. 2). Given this initial time investment, the Rand study further concludes that an additional ten to fifteen years of pioneer-plant operation is frequently required. Any hiatus in this twenty-year cycle will significantly diminish the potential of any improvements to reduce the process cost and will potentially limit the viability of the technology.

In a related study, Lieberman (1982) found that continuous and higher levels of R&D expenditures accelerate the rate of process improvement. Discontinuous R&D, it is speculated, has the effect of reducing the rate of process improvement using patents only in the area of coal liquids production. Figure 2–1 details the change in the patents rate after the 1974 and 1980 funding pattern changes and illustrates the drastic changes in patent productivity in these periods.

Regarding the overall effect of hiatus periods in the R&D process, Myers and Arguden (1984) have concluded that time, as opposed to people, facilities, or information storage systems, is the critical variable in determining the decay rate of process-oriented knowledge. Delays in the twenty-year cycle, they imply, would require the reinitiation of the development effort after each extended hiatus.

While the effects of hiatus periods on the production of basic knowledge are difficult to measure, it is known that funding-pattern reduction does affect the rate and direction of knowledge production (Crow and Bozeman, forthcoming; Link, 1982). Given this reduction in the rate of production and the fact that loss in the rate of knowledge production can result in a dissemination problem, it is speculated that periods of research hiatus result in excessive research duplication and reinitiation. These reinstitution efforts are often time consuming and lead to substantial delays in technology-development processes. Beyond the knowledge losses, there is the more general effect reduced funding has upon the search for potentially new processes that might be based on the findings of the basic research.

Figure 2–1
Patent Activity by Date of Patent Application in Vitro-Obtaining Liquid Hydrocarbons Processes (Patents Granted 1/67–12/82)

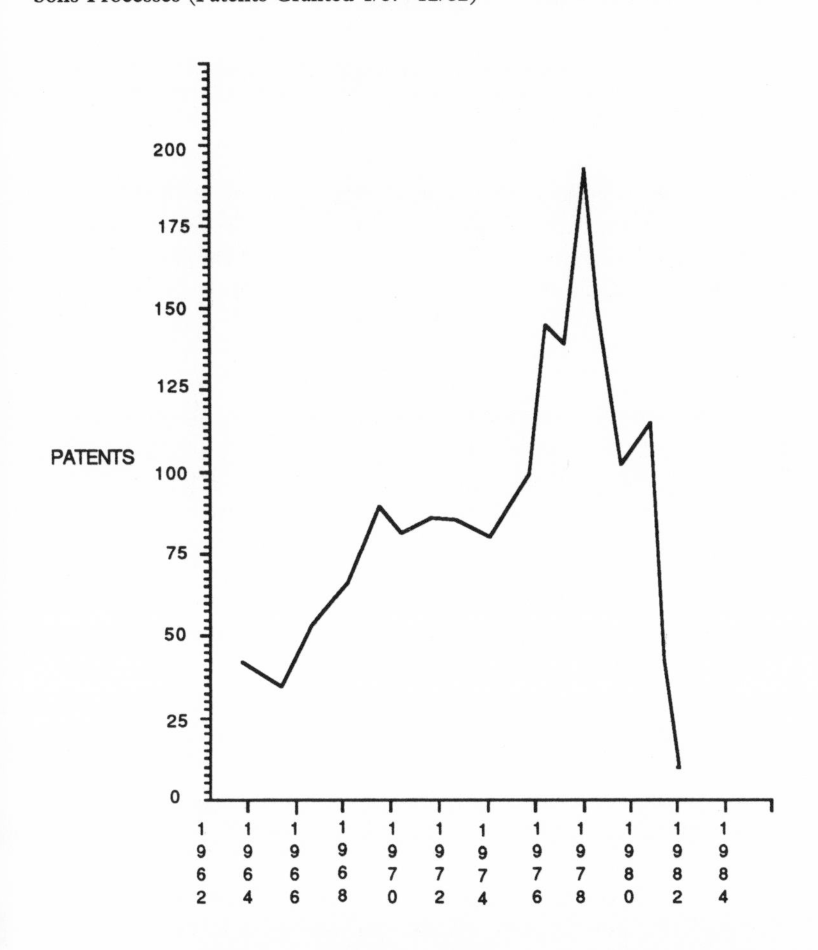

Sources: Patent Profiles, Synthetic Fuels, U.S. Department of Commerce, Patent and Trademark Office, December 1979; Index of Patents, U.S. Patent Office, 1979-1982;Official Gazette, U.S. Patent and Trademark Office, January 1979-December 18982

In sum, the hiatus effect on technology development reduces the ability of a process to develop along standard cost-reducing routes. The process-development cycle appears to take about twenty years for significant cost reductions to occur and for technical risk to be reduced to levels that will permit investment. Additional hiatus effects include a reduction in the rate of production of new knowledge and, thus, a reduction in the probability of developing new pioneer-process configurations.

SYNTHETIC FUELS TECHNOLOGY DEVELOPMENT AND THE HIATUS EFFECT

The development of SFT, particularly coal hydrogenation techniques, has been characterized by a mobilization strategy in response to a critical national need. As early as the 1920s, German coal hydrogenation RD&D efforts were carried out in great haste, almost in a crisis environment. Pier in 1925 obtained a good-quality gasoline from coal at the bench scale (Hughes, 1969). By September 1926 a plan to construct a 100,000-ton processing plant at Leuna had been developed and was announced (Donath, 1963). Completed after a monumental effort in 1927, the plant was fraught with technical difficulties and cost overruns. Production never reached more than 70,000 tons at the Leuna plant before 1929, and according to I. G. Farben, the developers, the plant was so expensive that more R&D was needed before moving ahead with any other plants (Hughes, 1969). Despite arguments to shut down Leuna and put a moratorium on R&D, development continued. This effort was carried out as a means of developing raw materials to aid the balance of payments problem Germany faced at the time.

Changes in the petroleum market between 1926 and 1930 forced the process developers to focus on development for internal markets and after 1933 on development for rearmament. Thus, the development of the eleven additional coal hydrogenation plants between 1936 and 1943 for the production of synthetic gasoline "flourished in an environment of rearmament and autarky" (Hughes, 1969, p. 130).

From a technology-development perspective, the German experience between 1927 and 1944 could be classified as a first-generation technology demonstration. Unfortunately, the plants were never intended to be competitive in the world market, and perhaps more importantly, they were not founded on a sound knowledge base. These German plants were also never fully developed in the sense that they lacked a systematic research-development and demonstration history over the twenty-year time frame suggested by Rand. Of course, they did operate, but very inefficiently and with numerous technical difficulties.

In the period between 1945 and 1948, American efforts to capitalize on the German first-generation SFT knowledge were launched in response to a perceived oil supply crisis (U.S. Congress, 1943). The plan, however, was based on several incorrect assumptions, not the least of which stemmed from the effect of the knowledge-loss rate in transferring the German technology base to the United States after the war. These included

1. the assumption that the German technology was sound because it had worked, when in fact it was not economically viable;
2. the assumption that the empirically developed German hydrogenation process could easily be converted to U.S. coal characteristics, when the variation in coal chemistry itself was the major cause of difficulty at Leuna and the other plants.

American policy in the 1940s had embarked on the path of demonstrating an unproven, uneconomical technology that lacked the science base necessary for improvements. The result was an unsuccessful four-year operation of the hydrogenation pilot plant in Louisiana, Missouri.

In the case of the Louisiana pilot plant, the hiatus effect began to have a significant impact on the SFT process. There had been only limited U.S. R&D on hydrogenation since the 1920s. Only three to four years of significant coal research before the beginning of the 1949 demonstration had been devoted to such work (Donath, 1963). Thus, the U.S. data base was limited and the German data base was spotty at best. Beyond the research base inadequacies, the pilot plant was only operated for four years before complete shutdown. According to the Rand studies, this would have been an insufficient time period in which to develop further the technology and to lower the costs. With the exception of some small-scale industrial efforts, the U.S. SFT development program entered a ten-year period of minimal research and development following the shutdown of the plant in 1953. The effect of this period of hiatus when the Office of Coal Research (OCR) began the development of three liquefaction demonstration projects in 1963 was a return of the technology-development process almost to square one.

Once again, as in 1929 with the Leuna plant and in 1949 with the U.S. project, the demonstration projects were overwhelmed by cost overruns and economic efficiency problems. By 1974, when the political cry for SFT development began in response to the Arab oil embargo, no OCR-supported demonstration project had yet been successful. While there are certainly numerous factors that contributed to the lack of success of these demonstration efforts, it is my contention that the principal reason for failure lay in the selling of these projects as demonstrations of a proven technology when in fact they were attempts to demonstrate a technology that was never designed to be economical and had already suffered two major periods of nondevelopment: the German–U.S. transfer after World War II and the 1953–1963 U.S. funding gap.

Out of a sense of panic and in an attempt to protect its bureaucratic status, OCR funded a major demonstration project in 1974. Known as Coalcon, the project suffered immediately from cost overruns and technical failure (GAO, 1977). In fact, it would not be unfair to conclude that the decision to proceed with the Coalcon project was a grievous error. It was already known that this SFT option was not ready for demonstration prior to embarking on the $237 million project (Desai and Crow, 1983). There existed no continuously de-

Table 2–1
Approximate Development Cycle for the Coal Liquefaction Option of Synthetic Fuels Technonogy

		Hiatus Period II
Program Shutdown	1985	Nearly complete reasearch, developmen t demonstration program shutdown
Premature Attempt to Commercialize SFT	1980 - 1985	Organnization and failure of the synthetic Fuels Corporation
Premature Development	1976 - 1980	Synthetic Fuels Commercial DemostrationProgram
	1974 - 1977	Coal con Demonstration failure
Premature Development	1963 - 1974	OCR develops and operates 3 liquefaction pilot plants (projects racked by funding and development problems)
	1961	Office of Coal Reasearch (OCR) established in the U. S. Department of the Interior. Small scale coal research/synthetic fuels research initiated
1951 - 1961		**Hiatus Period I**
	1952 - 1956	Carbide and Carbon Chemicals operates a 300 ton/day hydrogenation plant after 17 years of lab scale work
	1953	U. S. Pilot Plant in Louisiana, Mo. shut down after running six American coals. Cost studies show process not competitive.
Premature Development	1949	First U. S. hydrogenation pilot plant began operation in Louisiana, Mo.
	1945	New U. S. Bureau of Mines liquefaction lab erected at Bruceton, Pa.
	1941 -1944	12 million metric tons of synthetic liquids produced from first first-generation plants in Germany
	1936	U. S. Bureau of Mines initiates experimental work on coal liquefaction
	1936 - 1944	11first-generation plants built and operated by Germans
Premature Develpoment	1927 - 1936	Operation of Leuna Liquefaction Plant
	1927	First-generation hydrogenation demonstration plant built in Europe
	1920s	Germans discover 3 processes for coal conversion to liquid fuels
	1924	U. S. Bureau of Mines initiates small scale coal synthesis research

veloped data base and no stockpile of basic knowledge from which to make design decisions for the plant. The insufficient development time frame associated with the German technology and the lack of continuous R&D support resulting from the hiatus of 1953 to 1963, coupled with the inadequate development efforts of OCR between 1963 and 1974, doomed this project to failure from the start. The knowledge or technology base simply did not exist for any demonstration-plant efforts in 1974.

Given the hiatus periods in the SFT development history (see Table 2–1) and the time frame related to the successful development of process-oriented technologies, the 1974 Coalcon decision, the 1976 Synthetic Fuels Commercial Demonstration Program, and the 1980 Synthetic Fuels Corporation efforts could have been successful only if the U.S. program begun in 1949 had proceeded continuously so that by 1969 a second plant would have been operating for fifteen years. At this point, the knowledge might have existed for the development of a potentially cost-effective industry. As it was, the SFT process was plagued by periods of nondevelopment and, after 1974, the push for commercialization of the technology, not for the needed RD&D. The presence of the hiatus periods in the SFT development cycle made the commercialization goals set in 1976 and 1980 unobtainable.

In 1987 the U.S. SFT development process is once again in a period of hiatus. Once more, the knowledge developed in the past is being lost. When the SFT development cycle begins again sometime in the future, the nation will, once again, attempt to demonstrate a technology that in reality remains in its infancy. The successful development of SFT, as shown by Steinberg (1985), will require that the knowledge base be in place prior to any movement toward demonstration and commercialization of the projects.

CONCLUSION

The political decision to pursue the development of a national technology like SFT requires substantial public-sector and private-sector coordination as well as economic support for investment and development. This is particularly true for technologies that, unlike space and military developments, must compete within the framework of an existing market. These criteria are nothing new and are well integrated into administrative and financial decision making. What is missing from the political decision-making process is an assessment of the stage of development reached by a given technology prior to being embraced as a national objective. This point is not novel or original (Lambright, Crow and Shangraw, 1984), but in the case of SFT, it became a critical barrier to successful development.

It is difficult to determine where SFT would be today if the government had supported a national, stepwise RD&D program after it obtained the German technology base in 1945. It is also difficult to determine if SFT would have proven to be more viable had the initial German developers pursued the technology upon a stronger research base. Neither of these what-ifs can ever be evaluated. Sporadic government support and the premature attempts to commercialize SFT do, however, have a cost. That cost includes the billions spent on development to date, as well as the fact that we remain 20–25 years from being able to produce a science-based, developmentally sound SFT option that has any chance of operating at a level of efficiency that would merit commercialization.

This conclusion does not assume that the efforts in gasification and oil-shale technologies have yielded no technological advances. Indeed, there have been advances. What it does imply is that technology-development efforts have been shut down largely due to operating difficulties that in the end result in negative plant economics from a development perspective. For the most part, these operating difficulties are the result of process designs based on uncharacterized reactants, uncharacterized processes, and an uncharac terized product. The lack of knowledge, particularly scientific knowledge, is the explanation for these poor characterizations and poor process designs. Quite simply, the SFT development effort in the United States has failed due to the inability of the political system to invest in a long-term RD&D effort. This factor combined with a general societal overconfidence in the area of technology development is a road map for failure.

Unfortunately, the prospects for the future do not seem any better. The ultimate need for synthetic fuels is a certainty. Given the current hiatus in the SFT technology-development process and the improbability that any long-term oil supply interruption is likely to provide coal technology with a 20–25 year lead time, it can be safely concluded that highly inefficient SFT options will be our only means of meeting the future need. The result will be very expensive synthetic substitutes produced from inefficient plants. Thus, the real policy choice is between launching a realistic, long-term RD&D program 20–25 years before the need for low-cost synthetic fuels or proceeding with the technology developed by the Germans in the 1920s. In the absence of any attempt to eliminate the current hiatus, the SFT process will return to its earliest stages of development.

Perhaps the lead epigraph from the National Petroleum Council's 1953 final report might better have ended: "Therefore the NPC recommends that the government, working with industry, launch a systematic fuels research program to develop future synthetic liquid technologies." Regrettably, there is little political incentive to launch a 25-year RD&D effort that might be needed at some unknown period in the future. And with the price of gas going down, many shortsighted observers are predicting good times for the energy economy of the United States and smooth riding for the American consumer.

NOTES

1. The U.S. Synthetic Fuels Corporation was abolished on April 18, 1986, pursuant to the Consolidated Omnibus Budget Reconciliation Act of 1985 (Public Law 99–272). That law transferred the responsibility of monitoring current synthetic fuels projects to the Office of the Secretary of the Treasury. In addition, on April 19, 1986, the secretary of the treasury established the Office of Synthetic Fuels Projects to monitor technically and environmentally current federally funded synthetic fuels projects.

2. Research and development funding for synthetic fuels research dropped in constant 1974 dollars from a high in 1981 of $377.6 million to a low level of projected funding of less than $20 million total in fiscal year 1986. This funding level is the lowest annual federal commitment to research and development activities related to liquefaction and gasification of coal since the 1960s. Essentially, the funding level for the research and development activities sponsored by the federal government has been returned to the pre-energy crisis level.

3. Since 1982, the major synthetic fuels research operations at Gulf, Mobil, Exxon, Chevron, and other major energy corporations have essentially been shut down (per-

sonal communications between the author and research administrators in these companies).

4. For a summary of the four major thrusts see Crow and Hager (1985).

5. Macro-engineering is a term that is used to describe large-scale technological projects in terms of their technology and organization as well as in terms of their social and political dimensions. The American Society for Macro-Engineering is a recent outgrowth of the perceived need for the study and analysis of projects of this type.

6. For an excellent summary of the politics surrounding energy technology in general and synthetic fuels specifically, see W. A. Rosenbaum, *Energy, Politics, and Public Policy*, Washington, D.C.: Congressional Quarterly, 1981.

7. For a good summary of the ebb and flow of public policy surrounding synthetic fuels technology development, see Rudolph and Willis (1985).

8. I consider the creation of the Office of Coal Research during the Kennedy administration and the development of the Department of Energy in the Carter administration both to be examples of periods of increased central planning. In the Kennedy administration case, the increased central-planning tendency was a function of social programs, with the Office of Coal Research playing an important potential economic-development role in depressed areas of the United States, such as West Virginia. In the Carter administration case, central planning increased in that the government became the principal entity in developing a national energy policy.

9. After the end of World War II, the bulk of the money set aside for synthetic fuels technology development by the Congress was intended for demonstration-plant design and construction, as well as some demonstration-related research activities. Very little beyond the core basic research program, which had been going on in the Bureau of Mines since 1925, was funded. The same was true for the Office of Coal Research funding pattern as well as the synthetic-fuel technology-development funding patterns of ERDA and the U.S. Department of Energy. For an analysis of the funding patterns between 1971 and 1984, see National Science Foundation (1983). For funding patterns prior to 1971, the period 1961 through 1971 is covered by the annual reports of the Office of Coal Research. Those prior to 1965 must be found in the *Congressional Record*. Prior to 1961 funding patterns are identified in the Annual Reports of the U.S. Bureau of Mines.

10. Most of the actual development work carried out in synthetic fuels technology was implemented by industry under contract with the government. In most of these cases, the participating industry or group of companies would be responsible for the development or demonstration or pilot-plant facility. The government would support pursuit of this activity with often the lion's share of funding.

11. The pork barrel of the demonstration projects associated with synthetic fuels technology is well known and is well documented in Vietor (1984).

12. An excellent summary of the technology-development and demonstration efforts associated with the Apollo project and the details of the knowledge base known to the scientists involved at the time of the 1962 speech by President Kennedy is presented in McDougall (1985).

REFERENCES

Bilstein, R. E. 1980. *Stages to Saturn: A Technological History of the Apollo/Saturn Launch Vehicle*. Washington, D.C.: National Aeronautics and Space Administration.

Bozeman, B., and Rossini, F. 1979. "Technology Assessment and Political Decision Making." *Technological Forecasting and Social Change* 15:25–35.

Congressional Research Service. 1980a. *The Pros and Cons of a Crash Program to Commercialize Synfuels*. Washington, D.C.: Congressional Research Service.

———. 1980b. *Synfuels from Coal and the National Synfuels Production Program: Technical, Environmental, and Economic Aspects*. Washington, D.C.: Congressional Research Service.

———. March 1981. *Costs of Synthetic Fuels in Relation to Oil Prices*. Washington, D.C.: Congressional Research Service.

Cooper, Bernard R. 1981. *Chemistry and Physics of Coal Utilization—1980*. New York: American Institute of Physics.

Crow, M. M. December 1982. "Synthetic Fuels Development: The Impact of the Reagan Administration." Syracuse University Institute for Energy, Research Policy Series 82–3, working paper.

Crow, M. M., and Bozeman, B. Forthcoming. "R&D Laboratory Variation in the 1980s: Some Implications for Policy Analysts." *Journal of Policy Analysis and Management*.

Crow, M. M., and Hager, G. L. August 1985. "Political Versus Technical Risk Reduction and the Failure of U.S. Synthetic Fuel Development Efforts." *Policy Studies Review* 5(1): 145–52.

Desai, U., and Crow, M. M. August 1983. "Failures of Power and Intelligence in Government Decision Making." *Administration and Society* 5(2):185–206.

Donath, E. E. 1963. "Hydrogenation of Coal and Tar." In *Chemistry of Coal Utilization*, supplementary volume, ed. by H. H. Lowry. New York: John Wiley and Sons.

Energy Research and Development Administration. 1975. *A National Plan for Energy Research Development and Demonstration: Creating Energy Choices for the Future*. vol. 1, *Program Implementation*. ERDA 48, 2. Washington, D.C.: Government Printing Office.

Goodwin, C. D. Spring 1981. "The Lessons of History." *Wilson Quarterly* 5:91–97.

Harlan, J. K. 1982. *Starting with Synfuels*. Cambridge, Mass.: Ballinger.

Hess, R. W. March 1985a. *Potential Production Cost Benefit of Constructing and Operating First of a Kind Synthetic Fuel Plants*. Report N–2274–SFC. Santa Monica, Calif.: Rand Corporation.

———. March 1985b. *Review of Cost Improvement Literature with Emphasis on Synthetic Fuel Facilities and the Petroleum and Chemical Process Industries*. Report N–2273–SFC. Santa Monica, Calif.: Rand Corporation.

Horwitch, M. 1979. "Designing and Managing Large Scale Public-Private Technological Enterprises: A State of the Art Review." *Technology in Society* 1(1):179–92.

———. 1980. "Uncontrolled and Unfocused Growth: The U.S. Supersonic Transport (SST) and the Attempt to Synthesize Fuels from Coal."*Interdisciplinary Science Reviews* 5(3):231–44.

Hughes, T. P. August 1969. "Technological Momentum in History: Hydrogenation in Germany, 1898–1933." *Past and Present* 44:106–33.

International Energy Agency. 1982. *Coal Liquefaction: A Technology Review*. Paris: Organization for Economic Cooperation and Development.

Kash, D. E., and Rycroft, R. 1984. *U.S. Energy Policy: Crisis and Complacency.* Norman: University of Oklahoma Press.

Lambright, W. H.; Crow, M. M.; and Shangraw, R. May 1984. "National Projects in Civilian Technology." *Policy Studies Review* 3(3–4):453–59.

Landsberg, W. H., and Coda, M. J. February 1983. "Synfuels: Back to Basics." *Resources* 72:12–14.

Lee, Bernard S. 1982. *Synfuels from Coal.* New York: Aiche.

Lieberman, M. B. 1982. "The Learning Curve: Pricing and Market Structure in the Chemical Processing Industries." Ph.D. diss. Harvard University.

Link, A. N. 1982. "The Impact of Federal R&D Spending on Productivity."*IEEE Transactions on Engineering Management* EM–29: 166–69.

McDougall, W. A. 1985. *The Heavens and the Earth: A Political History of the Space Age.* New York: Basic Books.

Merrow, E. W.; Phillips, K. E.; and Myers, C. W. June 1981. *Understanding Growth and Performance Problems in Pioneer Process Plants.* Report R–2569–DOE. Santa Monica, Calif.: Rand Corporation.

Myers, C. W., and Arguden, R. Y. January 1984. *Capturing Pioneer Plant Experience: Implications for Synfuel Projects.* Report N–2063–SFC. Santa Monica, Calif.: Rand Corporation.

National Petroleum Council. February 26, 1953. *Final Report.* Washington, D.C.: U.S. Department of the Interior.

National Research Council. 1980. *Refining Synthetic Liquids from Coal and Shale: Final Report of the Panel on R&D Needs in Refining of Coal and Shale Liquids.* Energy Engineering Board, Assembly of Engineering. Washington, D.C.: National Academy Press.

National Science Foundation. 1983. *Federal R&D Funding for Energy Research in Fiscal Years 1971 to 1984.* Washington, D.C.: National Science Foundation.

Nelson, R. R. February–December 1959. "The Simple Economics of Basic Scientific Research." *Journal of Political Economy* 27:297.

Nelson, R. R., and Langlois, R. N. February 18, 1983. "Innovation Policy: Lessons from American History." *Science* 219:817–18.

Office of Coal Research. 1968. *Annual Report.* Washington, D.C.: Department of the Interior.

———. 1969. *Annual Report.* Washington, D.C.: Department of the Interior.

———. 1970. *Annual Report.* Washington, D.C.: Department of the Interior.

———. 1971. *Annual Report.* Washington, D.C.: Department of the Interior.

———. 1972. *Annual Report.* Washington, D.C.: Department of the Interior.

———. 1973. *Annual Report.* Washington, D.C.: Department of the Interior.

———. 1974. *Annual Report.* Washington, D.C.: Department of the Interior.

O'Hara, J. B. 1979. "Liquids from Coal." In *Coal Handbook*, ed. by R. A. Myers. New York: Marcel Dekker.

Penner, S. S. 1981. *Assessment of the Long-Term Research Needs for Coal Liquefaction Technologies (FERWG–11).* Washington, D.C.: U.S. Department of Energy.

Polaert, T. J. March 1985. "Status of German Coal Conversion Technology." *Chemical Economy and Engineering Review* 17(3):12–22.

Roland, A. 1985. *Model Research.* Vols. 1 and 2. Washington, D.C.: National Aeronautics and Space Administration.

Rothberg, P. F. September 1984. "SFT and National Synthetic Fuels Policy." *Congressional Research Service Review*, pp. 23–36.

Rudolph, J. R., and Willis, S. August 1985. "The Politics of Technology, Public Policy, and Administration: The Synthetic Fuels Venture in Western Democracies." Paper presented at the American Political Science Association annual meeting in New Orleans, Louisiana, 1985.

Sayles, L. R., and Chandler, M. K. 1971. *Managing Large Systems: Organizations for the Future*. New York: Harper and Row.

Science News. 1983. "The Dirty Face of Coal." *Science News*, September 17, 1983.

Seamans, R. C., and Ordway, F. I. 1977. "The Apollo Tradition: An Object Lesson for the Management of Large Scale Technological Endeavors." *Interdisciplinary Science Reviews* 2(4).

Shangraw, R. F. December 1982. *H-Coal: A Technology and Public Policy Cost Study*. Alfred P. Sloan Project Report, Syracuse University.

Stanfield, R. L. June 9, 1984. "Why Won't the Synfuels Corporation Work? The Real Problem May be Technology." *National Journal*, 16:1124–28.

Stares, P. B. 1985. *The Militarization of Space: U.S. Policy, 1945–84*. Ithaca, N.Y.: Cornell University Press.

Steinberg, G. M. 1985. "Comparing Technological Risks in Large Scale National Projects." *Policy Sciences*, pp. 80–93.

Technical Economics, Synfuels, and Coal Energy Symposium. February 1984. Presented at the Seventh Annual Energy-Sources Technology Conference and Exhibition, New Orleans, Louisana.

U. S. Congress. August 1943. *Hearings on Synthetic Liquid Fuels*. House Committee on Interstate and Foreign Commerce. Washington, D.C.: Government Printing Office.

Vick, G. K., and Epperly, W. R. July 1982. "Status of the Developments of EDS Coal Liquefaction."*Science* 217:4557.

Vietor, R.H.K. Spring 1980. "The Synthetic Liquid Fuels Program: Energy Politics in the Truman Era." *Business History Review* 54(1):1–34.

———. 1985. *Energy Policy in America since 1945*. New York: Cambridge University Press.

Wang, G.S.C. 1981. "Evolution of a Synfuels Project: An Engineer's Perspective." Proceedings of the IASTED Energy Symposia, Anaheim, California.

Whitehurst, D. D.; Mitchell, T. O.; and Malvina, F., 1980. *Coal Liquefaction: The Chemistry and Technology of Thermal Processes*. New York: Academic Press.

Part II

POLITICAL AND POLICY PERSPECTIVES

3

Technological Policy Making in Congress: The Creation of the U.S. Synthetic Fuels Corporation

PATRICK W. HAMLETT

A commonplace idea in modern policy analysis is that science and technology have come to play an ever more important role in our society. Dozens of recent books, articles, and commentaries dissect this development, analyzing and criticizing perceived relationships between past technological changes and contemporary social realities.

All branches of American government are involved in making and implementing policy decisions in technological areas. The contemporary literature, however, too often overlooks the role of the U.S. Congress in making technological policy, in large measure because of the attention paid to the more glamorous "technoscience" agencies of the executive branch—NASA, DOD, NIH, and others—and to the White House. Very dramatic technological developments and decisions surely emerge from within the executive branch, well deserving of attention and comment. However, Congress also makes many critical decisions on a variety of issues that influence science and technology or that have significant scientific and technological components. Energy, environmental protection, toxic substances, health, transportation, and R&D funding are within Congress' policy-making, appropriations, or oversight purview.

With a few notable exceptions, however, science and technology issues are, from a congressman's perspective, swallowed up into more traditional policy domains. In Randall Ripley's words,

> Members of Congress will not usually think of "Science Policy" as something either separate or sacred. Thus, if "Science Policy" has to do with the location of a costly government facility, such as an accelerator for atomic particles, members are likely to see the decision as another locational-distributive issue, not as a matter involving

the treatment and development of something abstract called "science." (Ripley, 1983, p. 415)

A study of the 1980 Energy Security Act is of interest here, not only for what it reveals of congressional procedures and politics in the sensitive arena of synthetic fuels, but also for what it displays of congressional handling of science and technology. In the end, the parliamentary tactics, coalition building, and bargaining and compromising in technological areas mirror the same processes at work in the nontechnical areas, a fact of some importance when we consider the widening significance of modern technology. Technological policies can be no better than the decision-making processes used to arrive at them.

THE ORIGINS OF H.R. 3930

The odyssey of synfuels legislation in Congress predates the 1980 Energy Security Act. There have been active proponents of federal support for synthetic fuels development since the 1940s, although they have functioned more as "voices crying in the wilderness" during the heyday of cheap and abundant natural petroleum. The problem with synfuels was not the feasibility of the technology, but production costs. Supporters of synfuels development thought their moment had finally come in the mid–1970s, as OPEC-driven world oil prices skyrocketed; they thought this would make synfuel prices competitive with natural oil. Their efforts came to naught, however, in 1975–1976, when a rather modest synfuels plan moved slowly through Congress. Despite strong support in the Senate, the bill died in the House of Representatives for a lack of a single supporting vote on a rules-suspending procedural matter. The bill never came up for a formal floor vote.

Two years later, staff members of the Subcommittee on Economic Stabilization of the House Committee on Banking, Finance, and Urban Affairs approached the chair, Congressman William S. Moorhead, with the idea of resurrecting the synfuels project. The staffers, Edwin (Ike) Webber and Norman Cornish, had recently visited South Africa, inspecting the SASOL I and SASOL II synfuels facilities there, and had what they thought was a new legislative strategy in mind for an American synfuels effort.

Webber, Cornish, and Moorhead, this time, were determined to so frame the new program that it would avoid the pitfalls of the 1975–1976 effort: unnecessary territorial fights between competing committees desiring jurisdiction over the legislation. They decided to draw the terms of the new project so narrowly that other committees could plausibly be denied jurisdiction altogether; this would prevent the program from being buried in other committees or amended beyond recognition by its opponents.

Fortunately, the perfect legislative vehicle for their purposes was close at hand. Shortly before, the Senate had passed and sent to the House S. 932, a simple, one-year extension of the 1950 Defense Production Act (DPA). As it turned out, the DPA fell entirely within the jurisdiction of Moorhead's subcommittee. They decided to attach their new synfuels program, as a substitute for the Senate version, to S. 932. Some quick calculations by Cornish revealed that annual military consumption of petroleum—an issue covered by the DPA—averaged about 500,000 barrels per day, which became precisely the production target for the new synfuels program. They felt there would be less opposition to a synfuels program aimed at freeing the military from dangerous dependence upon imported oil. Moreover, 500,000 barrels per day of synfuels production would be, they thought, sufficiently large to demonstrate the feasibility of commercializing synfuels technology. Thus, H.R. 3930, a substitute for S. 932, was born.

Moorhead's tactic of limiting the scope of the new program in order to end-run opposition in other committees worked, but not without a fight from Rep. John Dingell (D., Michigan). Webber feared that if Dingell's subcommittee got its hands on the program, Dingell would "gut the bill" (Webber, 1981). Moorhead, however, had designed the program so well that the House parliamentarian ruled in his favor, opening the door for rapid processing of H.R. 3930 by his subcommittee. On May 8, 1979, the Banking Committee reported H.R. 3930 to the House, recommending passage by a vote of 39–1.

The House took up H.R. 3930 on June 26, 1979. Majority Leader James Wright (D., Texas), who had favored the earlier attempt to formulate a synfuels program, thought the Moorhead program had considerable merit. Not only would the program get a federal synfuels program under way, but it would also demonstrate congressional—and especially House—leadership in energy matters. Democrats in Congress looked at the upcoming 1980 presidential election with considerable trepidation, in large measure because of the public's perception of interminable inaction on both the president's and Congress's part in solving the energy crisis. H.R. 3930 offered an opportunity to project firm, even dramatic, leadership on the part of the Democratic party. Indeed, the expected opposition from congressional Republicans could be depicted as foot-dragging when compared to the reinvigorated Democratic leadership.

Wright, however, thought the Moorhead program was too conservative, and he intended to introduce an amendment raising the production goal to 2 million barrels per day of synfuels, to be met by 1990 (Olson, 1981). Wright's amendment would clearly lift the program out of the limited scope of the DPA, and this caused some concern among the Economic Stabilization Subcommittee staffers, who had tried to carefully draft a limited program authority.

Prior to the floor vote on H.R. 3930, the House leadership—Wright, Speaker Thomas (Tip) O'Neill (D., Massachusetts), Moorhead, and others—met with President Carter, outlining the synfuels program as it was taking shape and seeking to gain his support. As Webber recalls the meeting, Wright informed Carter that the "synfuels train is leaving the station. Would he [Carter] be on board or not?" The House leaders served notice that the Congress would take the initiative on synfuels whether the president wanted it or not; Carter must join now, while he could still credibly claim a leadership role. Carter listened sympathetically, but made no commitment. Webber recalls an incident shortly after this meeting which illustrated the equivocal support for the synfuels program coming from the executive. A few days afterwards, Moorhead received a telephone call from Secretary of Energy James Schlesinger, who reported with a laugh that the White House was "onboard" the synfuels train. When Moorhead asked why the secretary chuckled, Schlesinger replied, "At least for the moment!" (Webber, 1981).

House opposition to H.R. 3930 on the GOP side of the aisle focused around Rep. David Stockman (R., Michigan) (later director of the Office of Management and Budget), Rep. Ron Paul (R., Texas), and around Dingell and Rep. Richard Ottinger (D., New York) among the Democrats. The Republicans argued that synfuels should be left to the marketplace, that the government had no business tinkering with free enterprise. Democrats worried that massive spending for synfuels would siphon off funds from other forms of energy production, and that the project as designed would offer, in Dingell's words, a "great opportunity for mis-management"(*Congressional Quarterly Weekly Reports*, June 9, 1979, p. 1099). With opposition building and the White House remaining lukewarm, Wright himself entered the struggle as floor leader for the bill. When, however, gasoline lines returned in the United States following the cutoff of Iranian oil, support for the Moorhead-Wright synfuels plan reached a fever pitch. Nothing, it would seem, would derail the "synfuels train" this time around, not even a congressional analysis indicating that in a "worst-case" scenario the program might end up costing a staggering $80 billion instead of the modest $3 billion contained in the bill (*Congressional Quarterly Weekly Reports*, June 30, 1979, p. 1286).

The bill's critics, sensing that the tide was against them, had virtually conceded the bill's passage. Instead, they attempted to add amendments to tighten up congressional control of the program. But the House was in no mood for such alterations, and all such amendments failed. On June 26, 1979, at the end of a long and sometimes rancorous day, the House overwhelmingly passed H.R. 3930 with Wright's amendment by a stunning 368–25 vote. Only the most steadfast opponents to the bill voted against it.

THE SENATE REACTION

President Carter, under intense pressure to "get moving" on the energy problems facing the nation, retreated to Camp David for the now-famous "Energy Summit." When he returned, he and his policy advisors had dramatically revamped their earlier energy program, shifting from a policy based on conservation to one based on increased production. Norman Cornish believes that it was the overwhelming House support for H.R. 3930 and for a federal synfuels program that triggered the Camp David retreat (1981). Whatever the case, Carter returned with a policy that now included an ambitious synfuels development program much larger than the House plan. Instead of the $3 billion program designed in H.R. 3930, the president's program would have diverted as much as $88 billion from the proceeds of a proposed "windfall profits" tax for synfuels development. Moreover, the president's plan called for the creation of a special government corporation, christened the Energy Security Corporation, to administer the money and the program.

Several members of the House leadership, including Moorhead, objected to the idea of a special corporation to administer the program. H.R. 3930 simply designated the president as administrator for the House plan, and Moorhead felt that a new bureaucratic structure would be unnecessary and counterproductive, duplicating existing administrative machinery. At one point, Moorhead joked that if a special corporation were needed, in addition to the new Department of Energy, perhaps Congress should create a special Defense Corporation to take over the job of the Pentagon (Moorhead, 1981).

Sen. Henry Jackson (D., Washington), then chair of the Senate Energy Committee and sponsor of one of a number of Senate synfuels bills, moved quickly to position his committee to process H.R. 3930. However, Sen. William Proxmire (D., Wisconsin), chair of the Senate Banking Committee, also wanted H.R. 3930 (now designated S. 932), since it had been reported by the counterpart House Banking Committee. Ultimately, the Senate leadership decided to assign the bill to both committees, and as a result, two different committee reports were prepared and submitted to the Senate for a vote.

Jackson's committee also carried the burden of processing Carter's synfuels program, although individual members of the committee—including Jackson—had reservations about the president's Energy Security Corporation concept. Senate opposition centered in the Banking Committee, led by Proxmire and Sen. William Armstrong (R., Colorado). Amstrong's opposition is significant because Colorado possesses the nation's largest shale-oil deposits and therefore would be one of the chief targets of any synfuels development program. Both Armstrong and Sen. Gary Hart (D., Colorado) were concerned about the social, economic, and environmental impacts of a major development program within their state and expressed their reservations openly. Armstrong, unburdened by questions of party loyalty on this issue, was strongest in his opposition: "Clearly, the proposal the administration presented to us today is a turkey" (*Congressional Quarterly Weekly Reports*, August 4, 1979, p. 1580). Hart's reluctance was more muted, described in terms of missed opportunities for other energy sources: "There are so many things we could do, that would be much less dramatic, with more immediate impact, and that would not cost billions of dollars" (Ibid.). Regardless of these objections, the members of the Banking Committee realized that congressional sentiment strongly favored some kind of program to develop synfuels. Their strategy, therefore, would be to craft a sensible alternative to the program emerging out of the Energy Committee and to offer the substitute to the Senate as a whole.

The program taking shape in Jackson's Energy Committee was an amalgamation of several proposals: the president's, the House version, Jackson's own plan, and a plan submitted by Sen. Pete Domenici (R., New Mexico). While neither the Jackson nor the Domenici plan contained explicit production goals, the president had set very ambitious production goals: 2.5 million barrels per day of oil substitutes by 1990 (U.S. House of Representatives, 1981). The Energy Committee was unwilling to accept all of the elements of Carter's program without alteration. Carter would have authorized the Energy Security Corporation (renamed the Synthetic Fuels Corporation by the committee) to invest $88 billion in development projects. The majority of the committee found this sort of "crash effort" in synfuels too ambitious. The committee restructured the president's program into a two-stage process, with the SFC authorized to provide up to $20 billion in financial aid over a three-year period. Congress would then reevaluate the entire program before authorizing the additional $68 billion. Congress would also impose stricter oversight controls on the SFC. "I think we have made a beginning," Jackson said, "but we shouldn't go off the deep end. The sensible thing is to start down the road with every possible safeguard, recognizing there are problems" (*Congressional Quarterly Weekly Reports*, August 4, 1979, p. 1580). The committee also delayed Carter's production target date to 1995.

The Banking Committee's plan was far less ambitious. Indeed, one irony of this history is that the Banking Committee's version of S. 932 rather closely resembled the original proposal that came from Moorhead's subcommittee in the House, which was the original core of S. 932. The Banking Committee rejected the fundamental premises of both the administration and the Energy Committee. Banking, somewhat reluctantly, wanted to promote synfuels development, but it wanted private industry to take the lead in doing so. A crash program, the committee feared, would tie the new industry to untried technologies and would inflict unknown environmental problems upon the Rocky Mountain states. Banking urged that a more modest $3 billion in financial help to private industry was wiser, both to find the most appropriate and survivable technology and to avoid the problems of more federal bureaucracy. No special corporation was proposed, and up to twelve different technical approaches would be aided. The financial mechanisms would be purchase agreements, loan agreements, or a combination of both. The Banking Committee's report concluded: "A crash program would be guaranteed to fail because it would attempt to do too much too soon." Failure would mean that the government would "be left with costly and inefficient 'white elephants' that would be less efficient, more environmentally offensive and more expensive to operate than future plants" that would come on line as the technology matured (*Congressional Quarterly Weekly Reports*, November 10, 1979, p. 2512).

Everyone knew that when both committees completed their mark-ups of the bill, the Senate would have to choose between diametrically different approaches to synfuels development. The Energy Committee leadership, Jackson, Domenici, and Sen. J. Bennett Johnston (D., Louisiana), thus began the task of lining up support for the Energy/administration version of S. 932. To win over skeptics and fence-sitters, Jackson and Johnston allowed the addition of other energy programs of interest to many members to the synfuels provisions of S. 932. In addition to the $20 billion that would be authorized for synfuels, the Energy Committee added $1.45 billion—through the Energy and Agriculture Departments—to encourage the production of alcohol and methane from agriculture crops and urban wastes, over $3 billion to establish a solar energy and conservation bank in the Department of Housing and Urban Development to subsidize loans for energy savings and solar retrofitting, and $85 million for a study of the acid rain problems of the Northeast, as well as instructions to the president to provide Congress annual energy and consumption targets through the year 2000 and to begin filling the Strategic Petroleum Reserve at a rate that would average 100,000 barrels per day. In effect, Jackson and Johnston added essentially unrelated energy programs that had active support throughout the Senate to the Energy Committee's synfuels plan. The effort swelled the $20 billion program to over $34 billion by the time the last provision was added. A single-goal energy plan had become a massive omnibus energy package in which nearly every pet project had found a place.

The two proposals came up for debate and vote from November 5 through November 8, 1979. The Carter administration had already made peace with the changes in the administration's plan demanded by the Energy Committee as well as with the numerous additional energy programs that had been included. The debate was long and often bitter. Sen. Proxmire repeated his concerns about the inflationary consequences of Energy's proposal. Jackson countered that the Budget Committee's subcommittee on synfuels, headed by Gary Hart, had found that the Energy version would have a minimal inflationary impact; he further asserted that the United States had "dawdled" long enough in launching a major synthetic fuels program. The Banking plan was "puny." "I suggest," he urged, "we send a strong signal, not a muted or confused signal, so that the world will know we are making a beginning" (ibid.).

Even traditional Democrats voiced concerns about the administration's plan. Sen. Adlai Stevenson (D., Illinois) suggested out loud that the Synthetic Fuels Corporation "has all the earmarks of a crash effort to get rid of an embarrassing $88 billion" from the proposed windfall profits tax (*Congres-*

sional Quarterly Weekly Reports, August 4, 1979, p. 1580). Sen. Howard Metzenbaum (D., Ohio) also expressed his doubts: "This piece of legislation could create one of the major boondoggles in the history of our country, or it could do what it is framed to do, which is to help us develop a synthetic fuels industry." He eventually sided with the Energy Committee because he thought that it "makes sense for us to take this calculated gamble" (*Congressional Quarterly Weekly Reports*, November 10, 1979, p. 2512).

In the end, the Energy Committee's version prevailed. Jackson and Johnston were able to keep their coalition together in large measure because they had sweetened the Energy Committee's version with enough plums to keep those who were only lukewarm about synfuels development on board. It is clear that the programs in gasohol production, solar credits, biomass conversion, and the strategic petroleum reserve, for instance, helped hold together the votes needed to defeat the Banking alternative. On November 8, 1979, the Senate passed the Energy Committee's version of S. 932 by a vote of 65–19.

Thus, S. 932 returned to the House, now a massive $34 billion omnibus energy package. What had begun as a modest energy end-run by the Economic Stabilization Subcommittee to provide for the military's petroleum needs through synthetic fuels had become something nearly unrecognizable to the House from which it had first emerged.

THE CONFERENCE

With the passage of the Jackson version of S. 932, the stage was set for the required House/Senate conference committee. Both chambers began selecting conferees. Three committees in the Senate were to send representatives: Energy, Banking, and Agriculture. Energy sent the largest contingent, since in many ways it was "their bill." Banking came along because of its earlier battle with Energy and because the Energy version of S. 932 contained matters that came under Banking's jurisdiction. Agriculture was represented because of the ambitious gasohol, biomass, and other renewable energy elements of the enlarged bill. Moreover, individual senators with special interests in just one title of the bill would be allowed to participate in the conference while that title was being considered.

Selecting House conferees was more difficult. Conferees from Moorhead's subcommittee had been selected as soon as H.R. 3930 had passed and while it was still assumed that the Senate Banking Committee would be the lead committee on the other side. Moorhead's strategy of narrowing the scope of his plan precluded the involvement of other House committees. But the bill that came back from the Senate had separate titles, besides its synfuels provisions, many of which had received no scrutiny by the House. Most of the added titles were clearly beyond the jurisdiction of Moorhead's contingent and arguably within the jurisdiction of many of the committees he had hoped to keep out of the process.

The conference presented the House leadership with a unique situation: unless S. 932 could be broken apart and the various elements handled separately, the conferees sent by the House would, in effect, be legislating for the entire chamber. With no corresponding House positions on several of the added programs, the leadership thought it wise to include on its conference panel representatives of those committees that would have dealt with such matters had they been submitted to the House according to usual procedures. Wright eventually chose conferees from four committees: Banking, Science and Technology, Interstate Commerce, and Agriculture. Wright made sure that the chair of each committee was among the conferees, in the hopes of lending authority to the conference's labors through their participation. The conference would have to operate outside usual House procedures, and Wright wanted to avoid potential challenges by having the various committee chairs already committed to the conference decisions.

The House eventually sent 23 members to the conference, swelling its ranks to 55 participants, plus 50 to 60 committee staff members from both chambers. Dealing with a bill of such complexity through so large a conference provided unusual logistic problems. Conference meetings ran with a formality of order and language not usually required from smaller groups and simpler legislation. It was decided early on that the various titles and provisions of S. 932 would be "farmed out" to smaller committees of Senate and House staffers according to their expertise. These "subcommittees" were to work conference decisions into legislative language, or, when this was not possible, to delineate areas of disagreement and potential solutions for conference review.

At the first meeting of the conference, Moorhead, the conference cochair along with Senator Bennett Johnston, suggested breaking off Title I, the synfuels part, from the rest of the bill. Since the House had already completed action on synfuels, but had no corresponding legislation for the other titles, this might allow the conference to complete at least one part of its job quickly. Johnston, however, quickly rejected any effort to sever the synfuels program from the rest of the bill, since Senate passage of S. 932 had been purchased only by the addition of those other programs. Johnston feared that without the support for those other programs, Senate approval of the conference report could not be assured. On the other hand, if the synfuels section could not be broken out, the House faced the difficult task of digesting the entire Senate version intact if it wanted to have any kind of synfuels legislation at all.

The specifics of the synfuels program formed the first hurdle facing the conferees: would they endorse the more modest House program or the ambitious and expensive Synthetic Fuels Corporation mandated by the Senate? To avoid months of wrangling as each house refought the philosophical battles that had occurred earlier in each chamber, Richard Olson, Wright's staff representative, suggested to Wright that the conference, in effect, do both plans at the same time. The two programs, Olson argued, were not mutually exclusive; they could be implemented as complementary parts of a larger program. The House DPA authority could be used, in Olson's words, as a "fast start" element of the program, getting government support to synfuels developers fast. The SFC, when finally staffed and operational, would then take over the management of the program, with the DPA authority reverting to a "stand-by" status, to be used by the president in times of emergency.

When Wright presented this option to Johnston and Moorhead, they both agreed to recommend it to their respective conferees at the first meeting. Wright's suggestion was adopted as the operating premise of the conference, partly because it offered a chance to avoid lengthy debate, and partly because it permitted each chamber to claim that its version of synfuels development had been adopted. This way, neither house had to concede to the other.[1] Not everyone was happy with Wright's proposal, however. John Dingell, who had led opposition to the original Moorhead bill in the House, was very uncertain of the "pork barrel" approach of the Wright proposal. To Dingell it seemed that simply deciding to enact both programs would provide opportunities for serious inconsistencies that would have to be ironed out. "Quick fixes" would not suffice to create a synfuels program with any hope of success (Ward, 1981).

Without exception, the staff members reported their work on the Energy Security Act to have been the most gruelling, most demanding, and most exhausting conference any of them had ever participated in.[2] Putting the conferees' ideas and compromises into formal language often required lengthy negotiation and debate among staffers representing the two chambers; working sessions often lasted well into the night and weekends.[3] Everyone labored under the burden of very complex legislation and the conferees' desire to complete the bill in time for President Carter to sign it on the upcoming Fourth of July.

When the conferees had completed their work, most of the Senate's synfuels provisions had survived. They agreed to the creation and staffing of the Synthetic Fuels Corporation and a two-stage development program with $20 billion authorized for the first stage. They agreed to the Government-Owned, Corporation-Operated (GOCO) provisions, but limited government-owned facilities to no more than three. They agreed to production goals of 500,000 barrels per day of synfuels by 1987, increasing to 2,000,000 barrels per day in 1992. The SFC would have a life span of twelve years and then would go out of existence. Its principal tools to induce private investment would be loans, loan guarantees, purchase agreements, price guarantees, and, if needed, the GOCOs.

With Title I very nearly complete, the conference turned to the complex titles dealing with conservation, solar energy, gasohol and methane production, acid rain studies, energy projections, and the Strategic Petroleum Reserve. Subcommittees of staffers had been hammering out package deals on various provisions. There were debates in each of these areas on both philosophical and practical grounds, but most staffers felt that these titles were handled with greater ease than the synfuels title had been.

As the conference moved into mid-June, there was a general expectation that the report would be completed quite soon, thereby meeting the leadership's July 4 deadline. Then, at the last moment—literally at the last meeting—John Dingell raised again his misgivings about the independent legal status of the SFC as a corporation: the committee wanted to create an organization outside the usual bureaucratic chain of command and accountability. Dingell, with the aid of his chief staff member, Michael Ward, objected that such a designation left too many unanswered questions and that the details of the Senate concept were so poorly worked out that it would virtually invite a flurry of court challenges. Dingell asked if the SFC would not be in violation of the Administrative Procedures Act, subject to conflict-of-interest charges, and beyond the restraints of the Environmental Protection Agency. In other words, Dingell asked if the Senate conception of the SFC as an independent corporation had not put it beyond the limits of a whole array of legislative enactments meant specifically to restrain government agencies.

Dingell and his staff felt from the start that the SFC provisions had been very poorly drafted, without the careful crafting and attention to detail for which Dingell himself had become well known (Ward, 1981). In effect, the challenge Dingell made to the conference was to force it to decide if the SFC should be a federal corporation, but exempted from any of the administrative laws that apply to federal agencies, or a private corporation, but subject to specific parts of the panoply of federal regulations.

Last-minute attempts were made to resolve the conflict, but the Dingell forces would not budge from their objections. Finally, Bennett Johnston submitted a series of amendments that removed, word by word, all references to the proposed corporations's private status. The Synthetic Fuels Corporation would be a federal agency, but exempt from the array of restraining regulations of other federal agencies; it would be, in Johnston's words, "*sui generis*."[4] Dingell was furious with what seemed to him to be an obvious ploy to paper over the technical problems he had raised, but Johnston's amendments passed each time over Dingell's futile objections.

When the seven-month conference completed its work, the report was returned for final approval in each house. The Senate passed the conference report on June 19, 1980, by a vote of 78–12. The report came before the House, where the odyssey had begun exactly one year earlier, on June 26, 1980. The first order of business was to gain a suspension of the House rules in order to bring this unusual piece of legislation to a formal vote.[5] It was the failure to secure a similar suspension of the rules that had doomed synfuels legislation two years earlier. This time, however, the procedural vote was a success, and the House proceeded to pass the conference report by a resounding 317–93 vote. The 1980 Energy Security Act, dramatically different from the much smaller and simpler original House bill, was ready for President Carter's signature four days ahead of schedule. On June 30, 1980, the bill was signed into law.

CONCLUSIONS

This history of the 1980 Energy Security Act prompts several observations. First, consider the magnitude of synfuels support that emerged in the 1980 bill. In the House, the original level of federal funding was a mere $2 billion, increased to $3 billion by the Wright amendment to the Moorhead bill. None of the original authors of the House verson anticipated its expansion to an unprecedented $20 billion (and quite possibly $88 billion). The Moorhead bill was modest in that the support levels were targeted only for defense needs and with the expectation that even a small demonstration program would draw private investments for a larger civilian program. The size of the Moorhead bill was also quite comparable to the earlier, defeated synfuels program of 1975. That the Senate Banking Committee would also come up with a federal support program of $3 billion strongly suggests that many members of Congress in both houses over a period of years favored a relatively restrained federal involvement in the development of this technology. Congress' own history in this field had never exhibited the huge levels of federal funding mandated in the Energy Security Act.

The strategy of coalition formulation in the Senate also reflects considerable congressional hesitation about so large a program. The addition of nine separate titles to the synfuels program was necessary to garner the votes needed to defeat the Banking Committee's alternative program. Bennett Johnston candidly admitted that the addition of the provisions for solar energy, conservation credits, gasohol production, an acid rain study, and so on was necessary to bring along members who were, at best, cool to the synfuels concept (*Congressional Quarterly Weekly Reports*, June 21, 1980, p. 1691). In conference, the Senate leadership insisted that the House deal with the entire Senate package, rather than splitting off the synfuels segment. Johnston explained that this was the only way to hold the Senate's coalition together. Clearly, much of the support for synfuels in the Senate relied upon political inducements rather than the technical specifics of the proposed synfuels project.

Second, the political environment of the synfuels program that emerged from the Senate appears in part to derive from the interchamber competition within Congress. Moorhead, for one, is convinced that the House action on H.R. 3930 caught the Senate off guard, spurring hurried efforts by Democratic leaders to reassert senatorial leadership in the face of the unexpected House preemption of the synfuels issue. The size and nature of the Synthetic

Fuels Corporation, as well as the other titles in the Act, Moorhead feels, reflected the Senate's desire to reassert its traditional preeminence in energy matters. To reestablish senatorial leadership, he argues, required more than simply aping the House's modest program. It required developing a dramatic and aggressive energy program that would put the House's effort in the shadows (Moorhead, 1981).

At that point, the Carter administration, facing severe reelection problems, joined and enlarged the Senate program in an effort, in turn, to reclaim energy leadership from Congress as a whole. The appearance of aggressive, creative energy policy making became a political prize sought after by each house of Congress and by the White House, in effect "bidding up" the programmatic and financial stakes of whatever energy program emerged in 1980. The House move caught the Senate by surprise, just as the emphatic congressional action caught the White House unprepared. None of the participants wanted simply to "me-too" the proposals of any other participant and thus the programs and budgets grew.

In the Senate, in particular, this meant adding to the synfuels program several other energy programs that had been languishing in the legislative process. Members of both chambers more enamored of solar energy or conservation than synthetic fuels development proved willing to trade support for synfuels in exchange for the inclusion of their favored alternatives. Too many smaller energy constituencies in Congress stood to gain by hitching to the Senate program, whatever their personal doubts or misgivings, to consider standing against whatever synfuels effort the Energy Committee wanted. The more modest Banking Committee alternative simply could not match the special-interest inducements of the Johnston-Jackson coalition. Once the bill emerged from the Senate, House members who wanted some kind of synfuels development program were forced to deal with the entire Senate program, whatever their private doubts, or face losing any hope of a synfuels program from this Congress.

Third, the pressures of a combined White House–Energy Committee campaign for the enlarged synfuels effort, plus all the other titles contained in S. 932, not only overwhelmed the House's (and apparently several senators') preference for a smaller synfuels program, it also overwhelmed the various technical analyses produced by private consultants and by congressional research agencies. These analyses almost uniformly disagreed with the optimistic production projections of the Energy Committee and the White House. The Senate Budget Committee, for instance, created a special subcommittee on synthetic fuels, headed by Colorado's Gary Hart. Hart required analyses of the Carter/Johnston/Jackson production goals as embodied in the bill emerging from the Energy Committee. Without exception, the congressional research agencies—the Congressional Research Service, the General Accounting Office, the Congressional Budget Office, and the Office of Technology Assessment—concluded that the production goals outlined in the bill were costly and unrealistic. Each agency preferred energy savings through conservation. Some synfuels development might contribute to solving the nation's energy problems, the agencies concluded, but political, technical, environmental, and financial considerations led them to conclude that massive synfuels development could not meet the targets then under consideration by the Energy Committee. The GAO report, for instance, asserted: "While synthetic fuels development is clearly an important and worthwhile national goal, we believe that conservation should take just as high or even higher priority" (*National Journal*, September 9, 1979, p. 1486). Bruce A. Pasternack, speaking for the private consulting firm of Booz, Allen, and Hamilton, Inc., testified before the Hart subcommittee that "a thoughtful half-million-barrel-a-day program" would be more advisable than the president's 1.75 million-barrel-per-day goal (*Congressional Quarterly Weekly Reports*, September 10, 1979, p. 1921). Hart's subcommittee eventually recommended that the government adopt a more modest program of testing competing synfuels technologies in order to find the approach most likely to survive on its own in the marketplace. Then the government would consider various supports to get the technology off the ground.

The reservations publicly expressed by citizen groups, environmentalists, and even some state governments about the potential ecological and social impacts of the enlarged synfuels program seem not to have influenced the congressional decision. Both Gary Hart and William Armstrong raised objections to the Senate's massive program, but without noticeable effect. Local concerns about waste disposal from synfuels projects, potential ground-water contamination by often carcinogenic wastes, the "boom-and-bust" cycle of resource development in rural communities, and increased demands for scarce western water needed in the synfuels process, while voiced by several groups in extended testimony before Congress, were unable to dampen the groundswell of support for a large-scale, "high-tech" energy program.

It seems clear that many members of Congress were willing to overlook the variety of long-term political and technical problems surrounding massive and rapid synfuels development in favor of various other energy programs contained in the Energy Security Act, or simply to appear to be taking decisive short-term steps to ease the gasoline shortages experienced by their constituents. From a partisan perspective, the Democrats expected to paint the inevitable Republican opposition as nay saying in a time that required bold, affirmative action. By including so many different energy programs within one omnibus bill, the Democrats were sure to reap the benefits of any successes that followed any of the programs.

Finally, the "high-tech" character of the synfuels program fit well with the American penchant for concentrated technological efforts to solve problems, as in the Manhattan Project, the Apollo moon mission, Project Independence, and the like. The American voter still holds technical expertise in high regard, and a highly visible technological campaign to correct the energy problem offered more immediate political payoffs than would a far more decentralized and diversified conservation effort, regardless of what such an effort might offer in real accomplishment. That members of both houses chose to ignore the bulk of technical analysis that expressed substantial reservations about the size, speed, and safety of the proposed synfuels development program supports the assertion of Senate Energy Committee staffer Daniel Dreyfus that

> almost by definition, the important issues facing the Congress are formed by national and world affairs. Critical and controversial political decisions, furthermore, can only be made when social pressures for decision are intense. Only then are signals from the polity adequate for evaluation of the political consequences, and only then would a prudent politician be compelled to make a hazardous choice. (Dreyfus, 1976, p. 270)

Technological decisions, like all others, must run the gamut of distinctly political concerns that structure a legislator's environment. In an environment of intense political pressures and a need for careful and acute sensitivity to the political—as opposed to the simply technical—consequences of each decision, it is not surprising that, in Charles Jones's words,

> for many congressmen, the purpose of policy analysis is to provide evidence for what their political judgment tells them is correct. Put another way, policy analysis in this setting is primarily a decision-supporting, not a decision-making process. (Jones, 1976, p. 263)

There was one surprise in this story. In disputes about technologies that represent radical changes within a concentrated social, political, and economic infrastructure, the expected dominant political pressures ought to oppose, rather than support, the proposed technical changes (Hamlett, 1983). Established relationships, investments, and jobs that must be affected by the changes under consideration are expected to form the basis for a coalition of opposition to the changes. Such a coalition did form in the synfuels issue, led by an unusual combination of usually hostile established oil companies and environmental groups. But the thrust of political pressures in the synfuels case was so strong in support of the new technology that opponents were outmatched throughout the process. Clearly, there is potential for counterintuitive results in the complex politics of congressional technology policy making, which argues for more extensive research into the political environment surrounding such technologies.

NOTES

1. Other staffers discussed slightly different versions of the origin of Wright's proposal (Olson, 1981).
2. Although the work was arduous, several staff members described the formation

of many "foxhole friendships" between House and Senate counterparts. Cornish and Webber spoke of an informal "S. 932 Club" of staff veterans who had worked together on this conference. Not everyone, however, reported good feelings. Michael Ward, who helped John Dingell stage his last-minute attack on the SFC, complained of attempts to isolate the House Interstate Commerce Committee staffers during the conference. Others, including Richard Olson, vigorously denied that any group had been isolated during the conference labors, that, in fact, extra efforts were made to hear all sides of each issue. There were also reports of hard feelings between Ward and Daniel Dreyfus, Senate Energy Committee staffer, concerning Ward's involvement in Dingell's last-day objections. Dreyfus described Ward as "destructive" and a "saboteur." Dreyfus's characterization of Ward, however, was explicitly denied by other staff members (Dreyfus, 1981; Cornish, 1981; Olson, 1981; Ward, 1981; Webber, 1981).

3. The debate among the conferees included detailed haggling about the specific powers, terms of office, and chain of accountability of the new Synthetic Fuels Corporation. Senate conferees wanted to make the corporation an independent agency free of the usual bureaucratic procedures and reporting requirements. House members, led by Dingell, wanted to impose strict congressional control over the corporation's decisions. Other issues included under what circumstances the House DPA authority would expire, when the president might reactivate that authority, exactly when the SFC could become "operational," what financial inducements it should use, and what relationship would exist between the SFC and the other federal energy agencies. One sticking point was the Senate's provision of so-called GOCO authority to the SFC. GOCO (Government-Owned, Corporation-Operated) authority meant that the government might own and then permit a private energy company to operate synthetic fuels plants. House members argued that this amounted to a governmental takeover of the new energy industry, while Senate conferees retorted that the GOCO authority was necessary to spur private corporations to invest in synfuels. Referring to the GOCO provisions as the "shotgun in the closet," Johnston argued that only if the private sector thought the government itself might enter into the industry would reluctant firms feel compelled to partake of the government-sponsored support contained in the Act.

4. Richard Olson characterized Johnton's amendments as intentionally "fuzzing" the status of the Corporation (1981).

5. Norman Cornish suggests that the House had to waive every rule "except the constitution" to get a direct vote on the conference report (1981).

REFERENCES

Congressional Quarterly Weekly Reports. June 9, 1979, p. 1099.

———. June 30, 1979, p. 1286.

———. August 4, 1979, p. 1580.

———. September 10, 1979, p. 1931.

———. November 10, 1979, p. 2512.

———. June 21, 1980, p. 1691.

Cornish, Norman. April 10, May 26–29, 1981. Personal interviews.

Dreyfus, Daniel A. 1976. "The Limitations of Policy Research in Congressional Decisionmaking." *Policy Studies Journal* 3:269–74.

———. April 16, 1981. Personal interview.

Hamlett, P. W. 1983. "A Typology of Technological Policymaking in the U.S. Congress." *Science, Technology and Human Values* 8:31–40.

Jones, Charles O. 1976. "Why Congress Can't Do Policy Analysis (or words to that effect)." *Policy Analysis* 2:251–64.

Moorhead, William J. April 14, 1981. Personal interview.

National Journal. September 5, 1979, p. 1486.

Olson, Richard. April 21, 1981. Personal interview.

Ripley, Randall. 1983. *Congress: Process and Policy*. 3d ed. New York: W. W. Norton.

U.S. House of Representatives. September 7, 1981. *Oversight: Synthetic Fuels*. Hearings before the Committee on Science and Technology, 96th Cong., 1st sess.

Ward, Michael. April 14, 1981. Personal interview.

Webber, Edwin. April 10, May 26–29, 1981. Personal interviews.

4

The Synthetic Fuels Corporation as an Organizational Failure in Policy Mobilization

SABRINA WILLIS

In June of 1980 Congress passed the Energy Security Act, Title I of which created the Synthetic Fuels Corporation (SFC). The intended goal was to encourage through financial aid the production of synthetic fuels in large quantities. Congress did not leave the SFC financially impoverished in assigning it this task. With $88 billion ultimately earmarked by Congress for the SFC, the development of synthetic fuels virtually became *the* national energy policy, and synfuels became so closely linked to the SFC in the public's mind and the political process that separating the two for purposes of analysis was rarely easy. What is clear is that the union neither benefited synfuels' political image nor worked to the advantage of developing synthetic technologies in the United States. More than six years after the creation of the SFC and half a year after its demise, the expenditures of the corporation have produced no new synthetic oil, but much controversy.

The problems of the Synthetic Fuels Corporation have been so many and so serious that it is difficult to determine exactly why the corporation failed to come close to achieving the goals set for it by Congress in the Energy Security Act. Journalistic articles concerning the corporation have used titles like "Liquify Synfuels," "Waiting for the Ax to Fall," "The Synfuels Fiasco," and finally, "Obituary for the Synfuels Corporation" (respectively, *Wall Street Journal*, February 1, 1981; *New York Times*, June 17, 1984; *Washington Post*, April 29, 1984). Public officials discussing the corporation commonly used words like "mismanagement," "cronyism," "ineffective," and even "the Bozo factor" ("Violation Cited," 1981). Proponents of the corporation and synfuels in general became few and their voices difficult to hear amid the cries denouncing the SFC's management, projects, and perceived extravagance.

Almost from the beginning the SFC had an image problem. In Congress the debate concerning the future of the SFC increasingly addressed not how much money should be appropriated for the synfuels venture from funds originally set aside for the corporation, but rather how best to reduce the size of the initial SFC allocation and then later how to scuttle the corporation altogether. The approach taken by Congress was to chip away at the SFC's funds and authority bit by bit. In 1984 Congress successfully rescinded $5.2 billion from the $14 billion the SFC had yet to spend of the $20 billion given it in 1980. This took place after $2 billion had already been diverted from the SFC in order to try to decrease the federal deficit (*Congressional Record*, September 26, 1984). As budget consciousness deepened and plentiful oil supplies began to look like a permanent condition, Congress and the president finished the surgery begun a year earlier. In the fall and winter of 1985 the corporation was relieved of its authority to make new financial commitments, and then its remaining $7.3 billion was rescinded. In December 1985 President Reagan signed into law a bill that mandated the closing of the United States Synthetic Fuels Corporation within 120 days.

SFC mismanagement and congressional attacks on the corporation explain much of the low esteem in which synfuels as well as the SFC have come to be held, but they do not explain either the origins of America's monumental commitment to synfuels in 1980 or all of the reasons behind the failure of the country subsequently to advance technologies whose time appeared to have finally arrived. To understand either of these matters requires at least some familiarity with the legislative origins of the SFC and a thorough discussion of the factors that influenced its operation after its creation.

INTEREST IN SYNTHETIC FUELS AND THE UNITED STATES GOVERNMENT

Early Interest

The idea of synthetic fuels technology and its noncommercial application have been around for a long time. In this country, government interest and support for research and development of synthetic fuels technology took place alongside private investment. Private industries, pursuing the very costly steps involved in developing and demonstrating the new technologies, accepted public funds in the form of loans, loan guarantees, and grants from the federal Bureau of Mines, Energy Research and Development Administration, and the Department of Energy, from states interested in synthetic fuels as a means of reviving the sagging economies of their coal mining regions, and later from the Synthetic Fuels Corporation.

The United States government first became interested in synthetic fuels to supply its modern, petroleum-fueled navy. In 1916, oil-shale land in Colorado and Utah was set aside to provide synthetic fuel in the event of some future petroleum shortfall. In 1926, under the direction of the Bureau of Mines, an oil-shale retorting plant was built near Rifle, Colorado (Vietor, 1980). Government and private interest in synfuels abated soon thereafter when new refining techniques and new oil discoveries ended (for the moment) fears of oil shortages.

Inspired by the German success with synthetic fuels and the American success with synthetic rubber during World War II, the American government renewed its interest in the development and production of synthetic fuels. Funds were allocated, primarily under the Synthetic Liquid Fuels Act of 1944. Again through the Bureau of Mines, pilot plants were built to examine direct and indirect methods of coal liquefaction. These remained in operation until the mid–1950s, when cheap Middle East oil discoveries made synfuels unattractive.

During the next decade and a half, interest in synthetic fuels in Congress waned but never died completely. Enthusiasts such as Congressmen Jennings Randolph and Carl Perkins, who authored the legislation of the 1940s and early 1950s, agitated periodically for federal involvement in the research and development of synfuels. However, given the long lead times and high investments required to produce synfuels in any great amount, it was not until the oil shocks of 1973 and 1979 increased the price of oil from under $3 a barrel to over $35 that the costs of developing a synthetic fuels industry again became politically acceptable. Bills designed to produce a small synthetic fuels industry were introduced in 1975 and 1976, but on both occasions failed for want of proper leadership (Hamlett, 1987). Then in 1978 Congress began work on its third serious attempt to create an American synthetic fuels industry.

The Creation of the SFC

The chronology of the SFC's creation has been well documented by Patrick Hamlett (1987). In 1978 the chairman of the House Subcommittee on Economic Stabilization, William S. Moorhead, hoped to get before the House a bill proposing the development of the capacity to produce the equivalent of a half million barrels of oil per day through synthetic-fuel technologies by amending an extension of the 1950 Defense Production Act that the Senate had already passed. The 500,000 barrels per day figure was selected because it was deemed large enough to demonstrate the feasibility of synthetic fuels production for commercial use and because it was approximately the amount of oil the military was consuming daily in 1978. Had Congress adopted the proposal as Moorhead conceived it, the country would have had a limited-scale synfuels project ensuring the development of pioneer demonstration plants for some technologies, plus an alternative to approximately 5 percent of the oil it was then importing from OPEC. The goal proved to be insufficiently ambitious when the full House considered the bill in the summer of 1979, several months after the fall of the shah of Iran and the rapid increase in the price of OPEC oil (Hamlett, 1987).

The subsequent trek of the bill through the legislative process is basically a tale of crisis decision making in which political actors in the House, Senate, and White House attempted to outbid one another in appropriating the synfuels idea as an answer to America's energy woes. By the time the bill passed the House (368–25) its original production goal of 500,000 barrels of oil per day (b/d) had been raised to 2 million b/d at the urging of the new House's Democratic leadership. Congress was congratulating itself for its bold leadership on the energy front and, as Hamlett notes, was inviting President Carter to "get on board" because "the synfuel train is leaving the station" (Hamlett, 1987).

Seemingly unprepared for the House passage of the bill, Carter's initial reaction was to hold an energy summit at Camp David and to try to set his own itinerary for the "synfuels train." Carter decided instead to commandeer the locomotive. His initial energy policy, with its emphasis on conservation and coal, was redefined to focus on the production of synthetic fuels. The president's program was also larger than the House version. The production goal was raised to 2.5 million b/d and the budget increased to $88 billion. Finally, the program was to be administered by an Energy Security Corporation and funded by the recently enacted windfall profits tax.

Consideration by the Senate did nothing to slow the bill's momentum, although it did shape its final form. The president's public corporation proposal, renamed the Synthetic Fuels Corporation, was retained as the basis for implementation of a synthetic fuels program in order to allow the body a measure of political independence and to give it a more businesslike character (Tierney, 1984). The Senate gave more time to turning Carter's plan into a two-stage process. During the first three years, the SFC was authorized to provide up to $20 billion in financial aid. At the end of this period, Congress would review the program and have the option of releasing the additional $68 billion. Otherwise, the Senate's final version was very close to the president's proposal. The date for realizing the 2.5 million b/d goal was, however, extended to 1995 and a $14 billion package was added to the bill to expand support for it by providing for creation of a strategic petroleum reserve, acid rain studies, the production of alcohol and methane from agricultural products and urban waste, and the establishment of a solar energy and conservation bank (Hamlett, 1987).

The Senate bill passed with a two-thirds majority at approximately the same time American hostages were being taken in the embassy at Teheran. Thereafter, the debate over the House and Senate versions essentially involved detailing the Senate version of the bill by giving the program a twelve-year life, setting final production goals of 500,000 barrels per day by 1987 and 2 million barrels per day by 1992, and providing the SFC with the right to aid synthetic-fuel projects through loans, loan guarantees, purchase agreements, price guarantees, and, if absolutely necessary, GOCO authority (Government-Owned, Corporation-Operated synthetic fuels plants run by private energy companies). The final version passed both houses ovewhelmingly, and by June 30, 1980, the United States had a new fuel policy: the promotion of a synthetic fuels industry, something quite different from what was intended by Moorhead's original proposal.

Reflections on the Act's Passage

The majority by which the final version was accepted in both houses would seem to indicate that, although the 1980 Energy Security Act was a complicated and bold piece of legislation, it was not severely challenged. Such was not at all the case. There was substantial criticism of the bill at every stage, from those who feared the inflationary effect of such a large expenditure to those who feared that the development and use of synthetic fuels would culminate in the melting of the polar ice caps (Nadir, 1979). Representatives of western coal and shale states expressed concern over the bill's environmental, economic, and social implications for their states, DOE proponents offered their fears that the SFC would usurp tasks rightfully belonging to that newly created department, still other congressmen expressed concern that the SFC would usurp rights and duties in energy development that should be left to the states or the Congress, and free marketers concentrated on the dangers of the GOCO approach to energy development (*Congressional Record*, November 7, 1979). Outside of Congress, oil lobbies, environmentalists, business interests, and some state governments opposed the version of the bill most akin to the act that finally passed. Finally, the analyses produced by the General Accounting Office, the Comptroller General, the Congressional Budget Office, the Office of Technology Assessment, and the Congressional Research Service for the most part expressed reservations about the untried nature of the synfuels technologies, and all concurred that the production goals were far too optimistic (U.S. Comptroller General, 1977; 1980; 1984).

These technological assessments were not, however, nearly as important to the policy-making process as the domestic concern with the recently doubled price of oil and the upcoming 1980 election. In such an atmosphere, the quick fix of synthetic fuels looked very attractive to a political process already geared to hard energy options involving advanced technology (for example, nuclear power). A businesslike, high-tech solution—one that required no sacrifice of the American standard of living but provided only the panacea of massive investment, magic corporate machinery, and wondrous technologies—appealed to a Congress and a president who were perceived as having nothing to offer in response to the oil shocks of 1979. When the SFC was being debated on the Senate floor, one senator reputedly said, "We have to do something even if it is the wrong thing" (cited in Stanfield, 1984), and another admitted that "frankly I am not particularly concerned whether the [appropriation to the SFC] number is $3 billion, $10 billion, $20 billion, or $88 billion, this Nation has to find alternative energy resources if we are going to be . . . a viable Nation in the future" (Howard Metzenbaum, in *Congressional Record*, November 7, 1979, p. 31185). This is precisely where the SFC's troubles have their origin: in its creation by a body with little understanding of the steps necessary to develop the technology of synfuels or of the unrealistic nature of the goals the SFC should be asked to fulfill.

THE SHORT, UNHAPPY HISTORY OF THE SFC

Whether the SFC could have managed to approximate the more realistic goal of 500,000 barrels per day had the gods been kind to it will never be known. From almost the outset the SFC's administration had three serious problems: the president, the economic environment in which it operated, and itself. The first two made it essential for the SFC to mobilize support for itself and its mission if it were to accomplish, even partially, the task assigned to it. The third largely made it impossible for the SFC to mobilize that support.

President Reagan and the SFC

Almost from its inception the SFC was an object of scorn for Ronald Reagan. As a candidate and as president he made no secret of the fact that he bore the corporation no particular affection. His principal appointee to the SFC Board of Directors was Edward Noble, a man who advised the Reagan transition team to disband the corporation. When Noble became the SFC's chairman, some believed that he was there merely to organize its collapse ("Synthetic Fuels," April 1981). While his view of the corporation later changed, President Reagan's did not. During the SFC's first year of life, the newly elected president neglected to make the appointments to the SFC's Board of Directors necessary to constitute a quorum and enable it to function. The SFC was also without a quorum and unable to conduct business for the better part of 1984 because the president refused to make the necessary appointments until Congress rescinded nearly half of the allocation remaining in SFC hands (*Congressional Record*, September 26, 1984, p. S11967). Thus, for nearly one-third of the corporation's total life span, President Reagan effectively barred the SFC from making any loans, loan guarantees, grants, or price guarantees by doing nothing.

Presidential hostility toward the SFC, however, was expressed more by passive resistance than aggressive activity. The White House vacillated for some time on the issue of allowing new commitments by the SFC. In the summer of 1985, for example, the Reagan administration supported Energy Secretary John S. Herrington in his denial of approval for the SFC to fund the Great Plains Gasification Plant in North Dakota, a DOE project. Subsequently the president gave Herrington permission to speak out against the corporation as a whole ("Synfuels Agency," 1985). Meanwhile, in August 1985, while Congress debated closing the SFC, the corporation's Board of Directors voted $744 million in subsidies for three synthetic fuels projects, only to decide in September to wait for a clear signal from the White House as to what course of action to follow. No clear signal ever came.

At the same time that the president was pressuring the SFC's Board of Directors to make these commitments, he also notified Congress that he wished it to liquidate the SFC. By the end of December, Congress had passed and the president had signed into law a bill rescinding the corporation's remaining $7.3 billion in funds and turning over its remaining administrative power to monitor those projects already funded by the SFC to the Department of Energy.

Throughout the fall of 1985 Herrington spoke out against allowing the SFC to commit any money to new projects to develop synthetic fuels. However, in December President Reagan privately urged the corporation to spend $484 million on two projects to extract oil from shale: a $300 million loan guarantee for Unocal Corporation at Parachute Creek, Colorado, and $184 million in loan and price guarantees at Seep Ridge. Though the SFC board was never offered any clearance in writing from the White House, the Treasury Department forwarded documents to the corporation that stated that the department had been consulted on the two projects, as mandated by law (for a summary of the projects of the California-based Unocal Corporation, see "Reagan is Urging," December 19, 1985). Because of these ambiguous responses, the SFC then decided not to fund either of these projects. On January 21, 1986, however, the SFC decided to make available $327 million in loan guarantees to Unocal Corporation. Rather than relying on congressional or administrative support, the corporation based its decision on legal opinions from SFC lawyers and from a Chicago law firm, both of which argued that since the board had a preexisting commitment to Unocal and Congress had exempted previous commitments from its ban on subsidies, the corporation was within its legal rights to grant the loan guarantee.

Market Factors and Synfuels

The other external factor that made life miserable for the SFC was the unpredictability of the world oil market and the downward drift of world oil prices after the SFC's creation. In 1980 world oil prices were at least ten times their pre–1973 levels. Energy forecasts were uniformly predicting that prices would continue to rise, with $100 per barrel prices not unlikely by the end of the 1980s. However, the worldwide recession and production disagreements between members of OPEC produced instead an oil glut that had depressed the price of oil to about $11 per barrel by the time of the SFC's termination.

Not surprisingly, this soft energy market affected not only public but also private investment in the development of synthetic fuels technologies. Indeed, during the period bracketed at one end by the oil shortages and soaring prices of 1973 and at the other by the oil glut and declining oil prices of the eighties, the most conspicuous synfuel projects tended to involve the efforts of the major petroleum companies to upgrade existing research and to demonstrate individually their pet synfuel processes at the demonstration-plant level inside the United States or for tar sands in adjacent Canada.

At times these projects involved public support. Ashland's Oil's H-Coal hydrogenation liquefaction project in Kentucky, for example, sought state and federal funding. However, for the major multinational oil companies' projects like Exxon's Colorado shale-oil project or Shell's Alsands project in Canada, public assistance in financing was not crucial. These companies possessed the capital for R&D investment during the windfall-profit days following the 1979 price increases by OPEC, and synfuels seemed a sound investment to companies projecting that the price of OPEC oil would be between $50 and $60 per barrel by the mid–1980s. (During this period, shale oil was projected as costing a maximum of $45 per barrel to develop and liquefied coal between $50 and $75 per barrel to produce.) Moreover, because technological breakthroughs in the production of synthetic fuels, especially via direct liquefaction techniques, were viewed as worth billions of dollars, these corporations jealously guarded their respective technologies, avoiding both cooperative partnerships with their principal rivals and federal funding of their research lest such funding lead to the dissemination of project results to their competitors.

The large oil firms also reacted negatively to the creation of a Synthetic Fuels Corporation on the grounds that it would primarily benefit their smaller, needier competitors. Of the country's largest oil companies, only Gulf agreed to participate in a project with 100 percent state financing. (It was only after the windfall profits tax began to cut into corporate profits that the attitude of American oil companies in general began to mellow towards the SFC as a means of recovering some of the lost profits.) This was the SCR-II coal liquefaction project slated for West Virginia and designed to be cooperatively financed by the United States DOE (50 percent) and by the governments of West Germany (25 percent) and Japan (25 percent). The project, which collapsed when Reagan cancelled DOE participation, and to which Gulf contributed approximately $100 million for site development, was one of the first large-scale synfuel projects to fold in the 1980s and the only major one to do so for essentially political reasons.

Elsewhere, economic considerations determined the fate of these ventures, and the recent history of large-scale synthetic fuels development by the private sector in the United States reveals some interesting points and some equally interesting gaps in information on the general subject of leaving national goals fundamentally to the private sector for implementation. The most important of these for our purposes and the SFC's is that private corporate actors are governed by corporate, that is, economic and commercial, criteria. Unfortunately for the cause of synfuels, the commercial competitiveness of synfuels by the early 1980s had been seriously undermined by the combination of inflation and high interest rates that increased the cost of producing them and by a global recession that resulted in a decline in the price of and demand for oil. One after another, the major petroleum corporations began to withdraw from synthetic fuels projects or to scale them back to pilot-plant proportions. Within a twelve-month period Exxon closed its multibillion-dollar shale-oil project in Colorado, and Shell, Cities Service, and Exxon withdrew from their tar sands projects in Canada, with losses involving hundreds of millions of dollars. Thus, the Synthetic Fuels Corporation had great difficulty by mid-decade in finding qualified, interested applicants for its assistance in synfuel demonstration projects.

Internal Difficulties

As damaging to the SFC as attacks by the Reagan administration and the changing nature of the oil market were, the corporation's worst enemy may have been itself: its questionable ethics, its inexperience, and the indecisiveness of its board of directors and staff. The corporation's first president, Atlanta real-estate developer Victor A. Schroeder, was forced to resign by 1983. His term in office had been continually marked by charges of bad judgment, but the final blow was based on charges of unethical behavior arising out of his offer to solicit business from the engineering firm of a member of the SFC's board of directors from whom he was seeking support for his plan to reorganize the SFC ("Synfuels President," August 9, 1983). It was six months before Oklahoma banker Victor Thompson was chosen to fill the vacancy, but only two weeks more before he, too, was in political trouble. In his case the problem was SFC criticism for banking mismanagement, which included charges of understating 1982 losses while serving as director of Tulsa's financially troubled Utica National Bank. The result was the same: for the second time in less than a year the SFC's president had to resign in disgrace ("Quit Synfuels Post," April 3, 1984). Matters were not a great deal better elsewhere in the SFC's upper management. Altogether, five of the corporation's directors and seven of its other top officials resigned between 1980 and 1985 out of frustration with the SFC's inability to execute its appointed tasks.

When the corporation did function, its actions were often open to valid attack, as in the case of the Santa Rosa, New Mexico, project. The SFC issued a letter of intent to fund this tar sands project, but later canceled it after the principal sponsor of the project withdrew. Later the SFC's oversight committee discovered that the letter of intent was based on incorrect and insufficiently analyzed information and should never have been issued in the first place (*Synthetic Fuels Oversight*, 1984, p. 15). Confirmation of the resource base was one of the specific conditions that the SFC's board of directors placed on the Santa Rosa project, but upon closer examination those reserves proved to be impossible to recover. The letter of intent promising SFC assistance did, however, enable project participants to sell $4 million worth of stock and effectively double the worth of their company (ibid., p. 17).

Finally, the SFC was criticized for the unprofessional-to-unethical practices of its officers. The list of charges grew throughout the corporation's brief life and included high-paying posts obtained by officers for their wives. Schroeder paid himself $135,000 and appointed his wife to a $45,000 position. Vice presidents were involved in conflicts of interest. Several members of the SFC's board retained substantial holdings in energy companies applying for SFC assistance. Corporate headquarters were lavishly refurbished for $500,000 when the directors moved in. Officers stayed at golf resorts on business trips, and one spent $30,000 in one year on junkets to Europe and South Africa. The SFC wasted almost $3 million in payments to consultants for tasks that might have been performed by its own small staff. Finally, the SFC allegedly thwarted the will of Congress by secreting allegedly sensitive, "proprietary" information pertaining to applications at its own "safe houses" and at the offices of applicants in order to keep it from Congress and, near the end of its life, systematically planned to commit funds before Congress could enact legislation to terminate the corporation. None of this SFC behavior was lost on Congress. Even supporters of the SFC had to admit, as Senator Metzenbaum concluded in defending a compromise (reduced) rescission bill in September 1984, that "nothing has been more damaging to the agency's credibility than the evidence and suggestion of impropriety on the part of the directors" (*Congressional Record*, September 26, 1984, p. S11963).

CONGRESS, TECHNOLOGY, AND THE SFC: THE MORE FUNDAMENTAL ROOTS OF THE SFC'S FAILURE

In short, it would be hard to imagine a more perfect prescription for a program's or agency's self-destruction than the SFC formula. Take an idea and race it through Congress, bloating it along the way. Assign its execution to a complicated corporate structure and staff the corporation with personnel who combine a dose or two of some of the worst factors occasionally associated with the Reagan administration. Add the outright hostility of OMB Director David Stockman and DOE Secretary John Herrington to presidential neglect of the agency and allow the resulting structure to settle in circumstances unfavorable to the pursuit of the goal it is charged with executing. Yet even from this perspective, the problems facing the development of synthetic fuels and the Synthetic Fuels Corporation were more fundamental and were symptomatic of an even larger defect: the decision to develop a synthetic fuels industry was made during a period of crisis and was based on a series of faulty assumptions and comparisons.

Since the 1940s, advocates have believed that a synthetic fuels technological breakthrough was always just around the corner, or even that the technology for synthetic energy independence was ready for application. In April of 1953, when the House was considering its appropriations for the development of synthetic fuels, Congressman Perkins paraphrased a report from the Interior Department Appropriations Bill of 1954 to the effect that "the coal to oil process has been proven to the point where the cost of production is within a few cents of the prevailing costs of securing gasoline and other refined products from petroleum." Even at this early date the contradiction in the government's support of synthetic fuels research and development was apparent to Perkins, who continued: "This has been urged as a reason for discontinuing the government research program and to let private industry take over the studies when and if they are ready" (*Congressional Record*, April 15, 1953, p. 3179). Although Perkins was wrong about the advanced state of synthetic fuels technologies in 1953, the paradox of government support for synthetic fuels development on the basis of commercial viability was plain. On the one hand, synthetic fuels technologies needed to prove themselves to be sufficiently competitive commercially to present an attractive alternative to imported oil to attract support. On the other hand, there was the belief that the United States government need not, in fact should not, interfere in an arena that could be safely left for the private sector to develop when the time was ripe. In other words, synfuels technologies, in order to receive continued public support, must always appear to be almost on line but not quite ready for commercialization.

Twenty-six years later, in 1979, the outlook was the same. Once again, a congressman called for public investment in synthetic fuels technologies in the style of a "new Manhattan Project." Congressman Samuel Stratton stated, in fact, "The technology is already there—in England, in Germany and in South Africa" (*Congressional Record*, July 10, 1979, p. HR7896). Once again synfuels were portrayed by policy makers as a technological solution that was just over the horizon and needed only to be financed through the initial investment period. As in the past, the attractiveness of the synfuels notion was proportional to the danger of the crisis facing the American way of life. While in the past the concern of the moment had been with the creation of a modern navy, then wartime security, the Soviet menace, and foreign trade imbalances, in 1979 the crisis was political blackmail from OPEC. And once again, as in the past, when the crisis subsided, government synthetic fuels projects were strictly curtailed or terminated altogether.

This feature of the American political process has been due, in part, to the fact that, just as the early Christian Gospels were told and later written in the belief that the beginning of the end was just over the horizon, energy scenarios have projected crises as continually deepening or as quickly recurring. When the crises have not deepened, but rather have passed and economic growth has resumed, public perception has changed. In 1955 the decision was made in Washington to eliminate all federal involvement in coal liquefaction and other alternative energy programs on the assumption that private industry would pick up the development of the technology. Under the Reagan administration the preference has also been to allow the private sector to take the lead in synfuels development and commercialization—an approach that once again takes for granted that the private sector is willing or able.

Synfuels proponents have consistently cited South Africa's success with synthetic fuels as a rationale for widespread American development of a synthetic fuels industry. It is not difficult to detect more than a hint of American chauvinism in such arguments: If South Africa can do it. . . . Unfortunately, the comparison is not appropriate at the level the argument is customarily pressed. The South Africans have been able to achieve their level of energy independence under totally different circumstances than those operating in the United States. Except for coal, and a grade of coal not especially suitable for export, the South Africans—unlike Americans who have a four-fuel economy and a variety of energy options—are energy-resource poor. More importantly, South Africa is not a consumer of energy in anywhere near the amounts consumed by the United States. The three SASOL plants produce 150,000 barrels per day, perhaps half of South Africa's oil needs. The 1980 Energy Security Act demanded the production of 2 million barrels of synfuels per day, 10 percent of America's daily oil use at the time of the Act's passage. Yet the number of American plants required and the costs of such synthetic fuels production have not been determined, since no project of such a scale has ever been tried. To believe that the technology is in place and that the social, political, environmental, and economic costs can be dealt with accordingly in the United States as they have been elsewhere is to base public policy on a series of unsupported assumptions.

In practice, the SFC compounded the problem by attempting to skip the demonstration steps necessary in the development of any new technology before attempting large-scale commercialization (Young, 1984). This approach caused many of the corporation's difficulties in finding projects to fund. Successful firms will not go into the production of synthetic fuels until they feel that the technology is in place. This is another reason why the companies traditionally involved in synfuels did not take part in the commercialization schemes promoted by the SFC but instead eliminated or drastically reduced their activities in the area and went "back to the lab." The companies that cooperated with the SFC, like Solve-Ex in the Santa Rosa project, usually had something to gain other than eventual profits from a synthetic fuels industry. The most aggressive promoters of the corporation were the construction companies that build the plants and the firms that market securities or sell equipment.

The SFC, even before being entirely stripped of its authority, was a policy failure. It is not surprising that the corporation was finally eliminated. Can synthetic fuels as a technology be rescued from the debris of the SFC? According to one widely accepted view of policy making, policy solutions never really die; they just fade from view for a time only to be later recoupled with new problems and new political events (Kingdon, 1984, pp. 73ff.). The idea of producing synthetic fuels from coal or shale has traditionally functioned as just such a policy solution. From 1916 to 1944 concern with synthetic fuels production was coupled with wartime security. During the cold war period, the fear was not of supply but of access to foreign-held reserves. The production of synthetic fuels in the United States was seen as a means toward that end. For the next eighteen years the synfuels issue languished, only to be returned to good health by the oil shocks of the 1970s and to be offered as a solution to a new policy problem: energy availability and high energy prices. Political events, an impending national election, and a national mood of anger and frustration opened what John Kingdon has called a policy window, and through it fell the Synthetic Fuels Corporation.

According to the same theory, once a problem is addressed, regardless of whether any solution is reached, policy makers tend to tire of it either because they feel that they have adequately dealt with the problem or to move on to other, newer, more glamorous problems (Kingdon, 1984, pp. 108–9). Hence, it might have been supposed that once created, the SFC would have remained unmolested and have had the opportunity to build its clientele under the cover of its bureaucracy. Such was not the case. Four years after its creation, the Congress began hacking the corporation down to a more manageable size, not as an admission by the policy makers of their errors in establishing the corporation or as an attempt to rectify them in order to better solve the earlier problem of energy security, but because, once again, policy coupling won out over meaningful policy making. The energy issue was forgotten in an effort to decrease the SFC's funds in order to meet the new menace: the growing budget deficits.

Although the SFC spent less than a third of its original, five-year appropriation during its existence, by 1984 the corporation had become connected with the federal budget deficit and the very symbol of government waste to its opponents. In June 1984 Congress rescinded $2 billion from its original budget, leaving the SFC with only $13.75 billion of its original appropriation and making it clear that the SFC had no hope of any future revenue from the $88 billion originally promised. In the minds of many congressmen, even the surviving amount was too much to leave the SFC, and three months later another $5.2 billion was rescinded. Unfortunately, Congress did nothing to reform the flaws in the corporation, nor did it even seriously attempt to refashion the SFC into an entity that could, if not increase energy stocks, at least increase knowledge of the processes for producing synthetic fuels. On the contrary, Congress chose within a year to totally dismantle the SFC.

In 1985 and 1986 Congress continued to divert money and authority from the corporation. Throughout the summer and fall of 1985 congressional critics like Mike Synar and John Dingell worked to prevent the corporation from committing any of its remaining funds to new projects. In December the SFC's remaining $7.3 billion was rescinded by Congress, and on April 18, 1986, the United States Synthetic Fuels Corporation closed its doors for the last time.

THE SFC AS AN ADMINISTRATIVE FAILURE: CONCLUDING COMMENTS

Congress, not the SFC, brought to an unsuccessful close the country's most recent attempt to develop synthetic-fuel technologies and to create a synthetic fuels industry. At present, the country still lacks a significant, ongoing demonstration project for any synthetic-fuel technology, excluding perhaps the Great Plains Gasification Project, and is not even considering anything beyond pilot plant-scale research in the liquid fuels, even though this was the core of the synfuels venture a decade ago. Thus, the government has once again gone out of the business of aiding the development of such large synfuel projects, although the experience of the last decade has shown that industry is unlikely to develop commercial-scale projects by itself until the feasibility of the technologies is demonstrated at the level of pioneer demonstration plants.

The question then is: What could the corporation itself have done differently to have been more effective in achieving the goals set for it? In 1980 Congress presumably chose to entrust the task of developing synfuel technologies to a public corporate structure, rather than leave the initiative for such work with the DOE, to ensure greater policy-making flexibility. Energy policy, when it has been made at all, has been very sensitive not only in the United States but in other countries that have tried to coordinate overall energy policies. Energy security and availability affect not only national security and economic productivity, but also individuals directly. Energy policy is, therefore, a highly political issue.

As a government corporation, the SFC was designed to have a measure of independence in choosing and funding its projects. From the beginning, however, its board of directors was dependent on the unofficial approval of the president to make any financial commitments. Congress, for its part, repeatedly accused the corporation of attempting to thwart its will. Like any new bureaucracy, the SFC needed to exert its independence. However, the corporation did not exercise its legal right to make a financial commitment over the strenuous congressional disapproval until after Congress had mandated its closure. Had the corporation earlier exerted itself for itself and wasted less of its political resources on attempts to please its nominal masters, it could possibly have done more, or, at the very least, have funded more projects that would have expanded our knowledge about the benefits and risks of developing a synthetic fuels industry. Barring this, the SFC might have been able to mobilize congressional and public support for itself. That the SFC was painfully inept at doing so became apparent when no group spoke on behalf of the corporation when Congress moved to disband it.

As a result, the SFC did more to harm the cause of synthetic fuels research and development than to help it. Although Ronald Reagan did a good deal to hinder the progress of the corporation, and market conditions and budgetary concerns were beyond the control of its board, the SFC's board of directors, when it had a quorum, did have some measure of autonomy. The SFC did not, however, press the limits of its authority or even question the directives of Congress and the president until after it had already been informed of its imminent closure. Had the corporation pursued an activist role earlier, it might, if nothing else, be remembered for more than the unprofessional and unethical behavior of its well-salaried employees.

Some of the blame for the final demise of the SFC must go to the corporation's Board of Directors and especially its chairman, Edward Noble. Although Noble did later change his original, negative attitude toward the corporation and even became one of its staunchest defenders, his managerial style was too weak to effectively guide a controversial organization. One observer noted that it was a Reagan strategy to make an appointment like that of Edward Noble to head the SFC, because it, like the EPA and other organizations unpopular with the president, were "patronage dumping grounds" where friends and important campaign contributors were sent (Ihara, 1986). The SFC still had the support of several powerful Republican senators until 1985, so President Reagan may not have been willing to invest the political capital necessary to demolish the SFC in 1981 or even 1983. Nonetheless, he could still effectively undercut the operation of the corporation during this period by delaying appointments and later by making appointees who may not have been chosen to be ineffective administrators, but of whom there was no reason to believe that they would be anything else.

The problem of Edward Noble's leadership was one of indecision. To say that the corporation was doing nothing would be unfair. After 1983 the SFC was, at least, awarding letters of intent and developing an overall strategy, but it was doing little or nothing to protect itself from the political attacks leveled against it. It has been said that not only was Noble "not an entrepreneurial decision maker," he was a political novice (Ihara, 1986). Because Noble was uncertain and inexperienced, he waited for cues from the Congress and the president on how he should proceed. Rather than attempting to shape or influence the political environment in which he was forced to maneuver or to take advantage of the corporation's nominal independence, Noble's approach was one of repeatedly reacting to events, a stark contrast to the more effective managerial style of the Carter appointee as SFC chairman, John Sawhill. In the few months that Sawhill ran the corporation, he attempted to mobilize support for the corporation among affected interests. He constructed a public-affairs unit to consult with state and local as well as national leaders. He attempted to forge ties with western state tar sands and oil-shale interests as well as those of eastern coal. He also made an effort to maintain good relations with the environmentalists. Noble, by contrast, neglected both in his contracts and his personal contacts eastern coal interests and the environmentalists, who later combined against the SFC (Ihara, 1986).

In conclusion, this country's failure to develop a synthetic fuels industry scarcely rests with the SFC alone. Even the best-organized agency staffed by capable people with spotless reputations would have had a difficult or impossible time finding worthwhile projects to fund in an environment where petroleum prices continually declined. At the same time, the SFC is the centerpiece of both a policy and an administrative failure in American policy making and is in good part responsible for the very low repute in which synthetic fuels are presently held. Today, synfuels have become synonymous with waste, fraud, and questionable business practices. We must, therefore, wonder if synthetic fuels as an idea can rise from the ashes of the SFC.

Past attempts by the federal government to foster the development of synthetic fuels technology indicate that it is an idea that will be resurrected. As in the past, the future will hold problems to which synthetic fuels production will seem the perfect answer. As John Kingdon noted, policy alternatives never really die. They just fade from view to recouple with new problems during later crises.

REFERENCES

Congressional Record. April 15, 1953. 99:3179.

———. July 10, 1979. 125:HR 7896.

———. November 7, 1979. 125:31184–31185.

———. September 26, 1984. 130: S11963, S11967.

Hamlett, Patrick W. "Technological Policy Making in Congress: The Creation of the United States Synthetic Fuels Corporation." This volume, Chapter 3.

Ihara, Randal H. September 18, 1986. Interview. Washington, D.C.

Kingdon, John. 1984. *Agendas, Alternatives, and Public Policies*. Boston: Little, Brown and Company.

Nadir, S. J. October 1979. "The Solution is the Problem." *Environment* 21:6–11.

New York Times. June 17, 1984.

"Quit Synfuels Post, Thompson Urged." April 3, 1984. *Tulsa Tribune*.

"Reagan is Urging Funding for Two Oil Shale Plants." December 19, 1985. *Wall Street Journal*.

Stanfield, Rochelle. June 9, 1984. "Why Won't the Synfuels Corporation Work? The Real Problem May Be Technology," *National Journal* 16:1124–28.

"Synfuels Agency Plans Criticized by Reagan Aide." October 3, 1985. *Wall Street Journal*.
"Synfuels President Under Fire." August 9, 1983. *Washington Post*.
Synthetic Fuels Oversight: Project Selection Procedures and the Santa Rosa Tar Sands Project. October 4, 1984. Committee on Government Operations Report. Washington, D.C.: Government Printing Office.
"Synthetic Fuels Were in the Spotlight—Then a Hearing." April 4, 1981. *Tulsa Tribune*.
Tierney, John T. Spring 1984. "Government Corporations and Managing the Public's Business." *Political Science Quarterly* 99:73–92.
U.S. Comptroller General. August 17, 1977. *First Attempt to Demonstrate a Synthetic Fossil Energy Technology—A Failure*. Washington, D.C.: General Accounting Office.
———. August 12, 1980. *Report to the Congress of the United States—Liquefying Coal for Future Energy Needs*. Washington, D.C.: General Accounting Office.
———. July 11, 1984. *Synthetic Fuels Corporation's Progress in Aiding Synthetic Fuels Development*. Washington, D.C.: General Accounting Office.
Vietor, Richard H. K. Spring 1980. "The Synthetic Liquid Fuels Program: Energy Politics in the Truman Era." *Business History Review* 54(1):1–34.
"Violation Cited at Bank of New Synfuels Chief." March 1, 1981. *Washington Post*.
Wall Street Journal. February 1, 1981.
Washington Post. April 29, 1984.
Young, Richard. November 26, 1984. Interview. Washington, D.C.

5

A State Government's Experience with the Synthetic Fuels Movement: The Case of Kentucky

JOHN C. MITCHELL

For over ten years the Commonwealth of Kentucky took part in a national synthetic fuels movement. The causes and history of this movement are linked to global politics, modern technology, American desire for both independence and big cars, and the U.S. economic growth of the early 1970s, among other factors. A comprehensive explanation of the attempt to industrialize synthetic fuels will not be discussed here. It should be understood, however, that such an effort followed the end of the era of secure and cheap oil and was envisioned as an important but partial solution to a national energy problem. For all intents and purposes, the movement is now stopped. Although a few large-scale projects proceed apace and various laboratories around the country continue to dabble in the technologies, the euphoria that once existed for the emerging synfuels industry is gone. It is appropriate to pause and assess what was done and why. For Kentuckians it is also important to understand what their state government undertook and what those activities were intended to accomplish.

THE SYNFUELS MOVEMENT

The oil shocks of 1973–1974 and 1978–1979 were both economic and political in nature. We now know that they were not temporary annoyances that Americans somehow had to deal with before returning to their normal way of life. Years later it was clear that they were precursors of a significant shift of economic power worldwide, a process resulting in the United States becoming a debtor nation whose once-great heavy industries, particularly steel and oil refining, were no longer making capital investments. No longer the champion competitor in the business of producing manufactured goods from a strong base of raw materials, the nation now seems resigned to grow only in the service and agricultural sectors of the economy. Thus, this occasion of "temporary" short supplies of crude oil signals not only the end of cheap energy, but apparently the backsliding of industrial growth in general.

If the oil shocks did nothing else, they forced the man on the street to realize that energy is a commodity not to be taken for granted. If a large supplier did not choose to do business with a customer who had high energy demands, then energy suddenly could be five times more expensive or indeed unavailable at any price. Moreover, it became obvious that there was precious little that government could do about it. Nevertheless, because the general public viewed shortages and high prices as unacceptable, the political system cast about for a government "fix" with little reliance upon the free market.

The strategy to achieve a quick-fix solution eventually centered upon industrial reform and took two forms to address both the demand and the supply of energy: conservation and the substitution of domestically produced coal for imported oil. While conservation caused concern among those who equated energy use with economic growth, it was regarded as "the most immediate opportunity" (Stobaugh and Yergin, 1979). On the other hand, increased coal utilization was an idea popular with those who favored a return to unlimited production based upon cheap, abundant energy sources as the path to economic salvation. More coal could be burned and, it was recognized, coal could be converted to more usable forms. Synthetic fuels then were a part, but only a part, of a national policy of achieving energy self-sufficiency.

The technology for liquefying coal as practiced in World War II Germany had received only minor attention from the U.S. government, taking the form of a few small pilot plants in Missouri in the early 1950s. Later in the mid–1950s a major synthetic fuels refinery based on coal was actually begun in South Africa. This refinery still operates, and two more coal refineries have been added to provide 80 percent of the country's needs, effectively insulating South Africa from Middle East oil politics ("Energy: A Special Report," 1981).

Coal-conversion technology development in the United States was not seriously launched until the early 1960s, when two predictions were made: one, that no more major oil discoveries would be made in the United States, and two, that making transportation fuels, even gasoline, from coal was technologically feasible and economically attractive. While neither prediction proved to be altogether true, they did serve to cause both formulation of technology-development programs by the federal government and acquisition of coal reserves by large oil companies. By the late 1960s, such companies as Exxon, Standard Oil of Indiana (now Amoco), Continental, Occidental Petroleum, and even Ashland Oil had acquired coal and created coal affiliates. Their action meant that more funds would be spent on coal-utilization research in the search for new markets by people who knew how to produce oil products.

At the time of the sudden energy disruption of 1973–1974, interest in coal technology was already high and coal conversion technology was well poised for large-scale development. The federal government reacted to the crisis by centralizing energy matters in a newly created Energy Research and Development Administration (ERDA). ERDA immediately decided that a new industry was needed to convert coal to oil and concluded that the best way to create this industry was to sponsor a cooperative program with industry to build large-scale demonstration plants. In a very short time several coal-conversion pilot and demonstration projects were proposed. In this atmosphere, the energy policy of Kentucky was born.

If the oil supply crisis of 1973–1974 precipitated the national synthetic fuels movement, that movement gained strength from the natural gas curtailments of 1976–1977 and the second oil shock of 1978–1979 closely related to the chaos in Iran. Dollar sales of energy to this country in 1977 were eleven times higher than before the oil embargo a mere four years earlier, yet the United States was importing more oil than ever before, 46 percent of total supply (Energy Information Administration, 1985). Worldwide inflation, in addition to control of oil production by foreign governments, eventually drove oil prices steadily upwards to record levels in 1981. Although a handful of pilot plants under ERDA and, after 1977, under U.S. Department of Energy (DOE) sponsorship were already operating or in the construction phase, it was reasoned that greater government spending to develop synthetic fuels was justifiable as one means of retarding price escalation.

In 1980 President Carter asked for and got the Energy Security Act that created the U.S. Synthetic Fuels Corporation (SFC) and authorized up to $88 billion for synfuels development. The SFC was to act as a banker, not a technology developer, to the emerging synfuels industry. Indeed, with oil prices approaching $40 per barrel in 1981, it seemed possible that a synthetic fuels industry could be established with minimum federal assistance. The cost-sharing formula decided upon was 25 percent in federal support (loan guarantees, price guarantees, price supports, direct loans, or combinations) coupled with 75 percent from the private sector.

Unfortunately, the concept never gained the full support of the Reagan administration. This meant that one of the three ingredients—executive branch, Congress, and industry—necessary for a successful program was missing. Then, due to a variety of factors, congressional support wavered. While some might argue that the federal policy was myopic, seeing only short-term problems and ignoring long-term worries, the energy companies themselves did not exhibit a particularly strong conviction that synthetic fuels were essential to their future. Over the course of four solicitations spanning three years, industry proposed too many projects that had either high capital costs, expensive products, tenuous technologies, uncertain financial and corporate backing, or questionable management behind them.

Of the 67 projects proposed under the first solicitation in 1981, only a few had carried design efforts far enough to maturity or had made even the minimum investment necessary to ensure a reasonable chance of successful scale-up. As industry rushed to join the synfuels stampede, too many technologies were proposed that would require substantial further investment and development work if cost overruns were not to occur. The result was a loss of credibility and a reduction of political support.

What happened to the production goals of the Energy Security Act, namely the 500,000 barrels per day of synfuels by 1987 and 2,000,000 barrels per day by 1992? The first answer is: the country did not need that amount of synfuels generated by $88 billion in taxpayer assistance. Conservation practices initiated in the private sector could achieve the energy-efficiency equivalent of that much production at no government expense. Second, building the number of plants for those production levels, about 40 plants for 2,000,000 barrels per day, would have created immense labor demands and imposed a severe strain on equipment suppliers in this country, not to mention the social, economic, and environmental consequences of such a building program.

At the insistence of Congress through the Deficit Reduction Act of 1984, the SFC pared back the program's aims from pursuing production goals to stimulating technical diversity. The intent of the limited program was to provide assistance to a few pioneer projects in the key resource categories of coal, oil shale, and tar sands so that "scale-up" risks would be removed from consideration when future commercial plants were needed. The Synthetic Fuels Corporation ultimately supported just four projects: one lignite, one western U.S. oil shale, one sub-bituminous coal, and one bituminous coal.

The national synfuels effort cannot be labeled successful. A better term would be incomplete. Admittedly, a few projects were assisted that may someday provide industry with important experience for enhancing the country's energy security. The country continues to depend on fossil fuels and our domestic oil resources are being depleted, but the American people do not seem ready to return to the horse-and-buggy mode of transportation. So, it is doubtful that sufficient knowledge was gained in the last wave to serve our long-term needs.

KENTUCKY'S POLICY

Kentucky's state government began to actively support the coal industry in 1972. The General Assembly passed the severance tax on coal that year, not only as a means to increase state revenues as more coal was sold, but also to appease coal-producing counties who clamored for the tax as a way to help maintain and improve their infrastructures. Although the coal producers accepted the tax graciously, they made it clear that in return they expected state government to work diligently to keep the coal industry healthy. A research and development program was thus established whose stated purpose was to find new and better ways to utilize Kentucky coal.

The major component of this R&D program of the Kentucky Energy Cabinet is housed today at the Kentucky Center for Energy Research Laboratory near Lexington.

In 1974 the General Assembly added another program having basically the same objectives, but much greater scope. The Energy Development and Demonstration Trust Fund was created at the request of Governor Wendell Ford, with urgings from development-minded state government and university officials and the advice of energy companies in Kentucky. In view of surpluses in the state budget existing at the time, government leaders agreed to commit up to $50 million for state participation in large-scale projects. In 1978 the limit was raised to $55 million. The specific purpose of the program was to allow the commonwealth to participate in large-scale pilot and demonstration projects in order to "encourage and promote the development and demonstration of efficient, environmentally acceptable, and commercially feasible technologies, techniques, and processes for conversion of coal into clean fuels in liquid, gaseous, or solid forms" (Kentucky Revised Statutes, 152.750).

Although these two programs were unique among the states, there is nothing extraordinary in the fact that Kentucky created them. One stimulus was the aforementioned budget surpluses, generated in part by the coal industry itself. Many public officials asked why resources should not be used to further economic development, long recognized as a proper function of state government. Then there was the economic fact that approximately 50,000 Kentuckians were employed in the coal industry and many more in several related industries. In addition, the federal energy agency was touting a national strategy that was expected to result in large-scale pioneer plants dotting the country and producing oil and gas substitutes from advanced technologies.

Heightening state interests was the fact that Kentucky was the leading coal-producing state and that, according to the Carter administration, coal production was going to double by 1985. It was reasoned that by playing an active role, Kentucky might gain more than its share of the federal synfuels budget. Furthermore, from plants located in Kentucky higher-valued products would pour that, in addition to increased sales of steam coal, would benefit Kentucky from the value added to coal still inside its boundaries. Finally, members of Kentucky's academic community had pointed out in the early 1970s that the body of expertise in advanced coal technology at that time resided beyond the borders of the nation's most productive coal state. By causing research to be carried out in Kentucky, it was felt that the state's image would be enhanced as it made a bid to become a leader in coal research and development. Along with recognition of this expertise, so it was argued, would come investments. Was this not shown in Silicon Valley, where money and jobs followed the high-tech bandwagon? Was not North Carolina successful in attracting investments in conjunction with its government- and university-supported complex at Research Triangle Park?

Kentucky's participation in the synthetic fuels movement was not only visionary and aggressive; it was well intentioned. The mechanisms put in place by the policies of the early seventies are, to a degree, still at work, although those policies have had to be modified to fit the times and today's economic realities. How does Kentucky's balance sheet on energy look after fourteen years?

KENTUCKY'S PROGRAMS

By 1988 Kentucky will have spent over $73 million on laboratory research. Although the funds have been used for capital outlays for building and equipment as well as research projects, the R&D program has been operated since 1972 primarily to support synthetic fuels. While current state outlays in this area are much lower, the laboratory continues to carry out research on liquefaction and oil shale and is considered a leader in certain technical areas such as catalysts, solvents, materials, coal preparation, and combustion. The R&D program will not be discussed here in detail since it was only an adjunct to the synthetic fuels movement. Rather, our focus will be the specific program aimed at encouraging large-scale production of synthetic fuels, the Demonstration Program.

From 1974 through 1985 the Commonwealth of Kentucky expended $27.4 million for seventeen large-scale projects (Kentucky Energy Cabinet, 1986). In capacity these projects ranged from "commercial size" (over 20,000 tons per day of coal and up to 50,000 tons per day of oil shale) to "demonstration size" (up to 6,000 tons per day of coal), and included pilot plants (up to 600 tons per day). Among the technologies represented were coal liquefaction, high-BTU gasification (pipeline-quality gas), medium-BTU gas, and low-BTU gas, as well as coal and oil-shale pyrolysis and coal-oil mixtures. The greatest share of state expenditure by far went to liquefaction projects in eastern and western Kentucky.

In fiscal year 1981 four large synthetic fuel projects were being developed for western Kentucky, as shown in Table 5–1. The table shows the very high levels of capital investment in the projects, most of which would have been in Kentucky, and the increase in coal production required to supply the plants. Not only would the economic activity have been enormous, but the 85,300 tons of coal feed per day, or about 28 million tons per year, would have represented roughly a 20 percent increase in annual Kentucky coal production and approximately 4,000 to 6,000 mining jobs. Output from the plants would have totaled more than 170,000 barrels per day in highly valued products. State government expenditures totalling nearly 16 million dollars for these four projects would have created almost 4,000 permanent jobs when the plants became operational. Naturally, the prospect of building these plants generated considerable concern about the region's ability to accommodate the projects within the existing infrastructure. Contingency planning

Table 5–1
Synfuel Projects Proposed in Kentucky, 1981

Project	Estimated Cost (billions)	Capacity (ton per day of coal)	Ky Funds Expended (millions)	Potential Employment
Breckinridge	$3.2	22,500	$4.2	1,300
W. R. Grace	3.0	28,800	0.9	532
Tri-State	4.0	28,000	5.0	1,200
SRC-1	2.2	6,000	5.7	757
Totals	$12.4	85,300	$15.8	3,789

was carried for the worst case where all four plants proceeded as scheduled and where the construction manpower plans of all four overlapped (Kentucky Energy Cabinet, 1981). However, there was little likelihood that all four projects would be constructed. Since the projects were never built, most state funding was used for design, operation, and pilot-plant test runs on Kentucky coal. A portion, almost $7 million, was used to acquire access to the proposed sites under option or purchase agreements.

Over the course of the demonstration program, state funds were expended on projects representing a total potential investment of over $14.7 billion. If all seventeen projects had come to fruition they would have resulted in an increase of some 47 million tons per year of coal and oil-shale mining and an increase of approximately 5,400 jobs in direct employment. A study performed by the Kentucky Energy Cabinet in 1983 showed that one commercial-scale coal liquefaction plant, Breckinridge, utilizing 22,500 tons per day of coal to produce 50,000 barrels per day of synthetic crude oil, would have provided significant economic benefits to Kentucky, particularly in terms of employment, increased economic activity, and additional state tax revenues (Holmes, 1983a).

To illustrate, direct employment of Kentucky residents during construction of the plant would total 4,000 persons at the time of peak activity; if the indirect effects were included, the total number of jobs would rise to 7,000. Income in Kentucky associated with direct and indirect plant construction employment would grow to $815 million. In-state purchases of goods and services for plant construction would be $1.3 billion and, with induced expenditures added, would be responsible for the production of $2.3 billion worth of goods and services in the state.

Plant operations following construction would directly employ approximately 1,300 Kentucky residents. Total direct and induced employment from plant operations and in-state coal purchases might have amounted to almost 4,900 permanent jobs. Income associated with this employment could total

$160 million annually. In-state purchases of goods and services when the plant was operating would approximate $302 million annually and would jump to a total of $544 million worth of goods and services in Kentucky annually if induced spending was incorporated. Direct and induced state tax revenues from plant construction were estimated at $94 million, while during each year of operation, tax revenues to the state could be as high as $32 million. In other words, the tax revenues generated by only one commercial-scale project in a single year would have been more than the cost of the entire ten-year demonstration program. Given the returns on only one project, there is little doubt why Kentucky officials were willing to take the gamble.

In actuality, only one project was completely constructed and put into operation: the H-Coal pilot plant project of Catlettsburg. State tax revenues generated by this project were estimated at $5 million, whereas state funds invested in the project amounted to $6.7 million (Holmes, 1983b). Employment peaked at 830 people during construction, and a full-time staff of 240 persons was employed during the two and a half years of operations. Investment in that plant by all sponsors totaled $323 million.

In addition to the $27 million spent by the state on all these projects, approximately $79 million was spent by the private sector and approximately $455 million by the federal government. Therefore, a total of $572 million was expended on the synthetic fuels movement in Kentucky. Of what benefit was all of this activity to the state?

THE OUTCOME OF THE MOVEMENT

It is difficult to quantify benefits from research. If the objective of Kentucky's program was to create an industry, then unquestionably the program was a failure. This is the case even though the H-Coal pilot plant, the largest direct liquefaction plant ever built in this country, was successfully completed, because it was not intended to produce a saleable product. But were there no economic benefits derived from this massive state effort?

One way to determine the economic benefits would be to figure the number of dollars spent in the state. Based on the data from a project-specific study conducted in 1983 by the Kentucky Energy Cabinet and additional data obtained from sponsors of other projects, it is estimated that of the $572 million spent on all projects, between $152 million and $218 million was spent in Kentucky for labor, goods, and services. Whether the state was paid for its $27 million investment is difficult to say.

One can also account for benefits elsewhere. For example, the energy research laboratory in Lexington is recognized as a leader in coal and oil-shale expertise, judging by the number of countries that have supplied visiting scientists: China, Poland, Israel, and Hungary. The laboratory continues to attract industrial and federal support, although not enough to reduce significantly its role as a state lab.

In addition, the synfuels movement has spawned two useful by-products completely unrelated to technology or dollars. One is the Cancer Register, the result of an epidemiological study conducted statewide to monitor the possible health effects of synthetic fuels. Another is the Population Distribution Impact Model, which predicts shifts and migrations in worker populations due to large projects. Developed by the Urban Studies Center at the University of Louisville specifically for the proposed synfuels industry, the model will be used to study the impact of the Toyota plant presently under construction in Scott County, Kentucky.

Major advances stemming from the synfuel program in utilization of Kentucky resources include the following:

1. Kentucky coal was run in the world's largest gasifier at the world's largest synfuels complex in South Africa.
2. Texaco gasifiers in Tennessee and Alabama, similar to those in use at the highly successful Cool Water electric power project in California, consistently run on Kentucky coal.
3. The main body of knowledge surrounding the ten-year development of Solvent Refined Coal is today based predominantly upon Kentucky coal.
4. The utility of Kentucky coal for hydrogasification, which is the most advanced method for making pipeline-quality gas, has been verified.
5. Kentucky oil shale has been tested in Brazil at the largest and only commercial oil-shale retorting facility in the world.

On a grander scale, history will record that the movement brought coal-conversion technologies from their moribund state after World War II to a new level of technical advance in the 1980s. Modern engineering knowledge and techniques were applied to achieve greater design sophistication, more certain methods, better controls, more reliable equipment, higher efficiencies, and, above all, a cleaner environment for the plant worker and the public. Processes that had been developed to refine heavy crude oils were successfully applied to coal. Through the use of modern catalysts, processes were improved and operated at significantly lower pressures than ever before. Gasifiers developed to produce synthesis gas for the chemical industry were modified and are running on coal today.

Whether the technical achievements have been worth the cost of the short-lived and relatively unproductive synthetic fuels program is a question that cannot be answered until oil is again in short supply. Thus, synfuels remain a good idea for the nation, for they are an assured energy source. Synfuels remain a good idea for Kentucky, for they could lead to large economic benefits. At the very least, one can take solace in the words of Edward Noble, the outgoing chairman of the Synthetic Fuels Corporation, who, in looking ahead to the next oil crisis, said, "If we had done nothing, there wouldn't be any alternative but another big government program. And this may save another big government program" ("Interview with Edward Noble," 1986).

CONCLUSION

As envisioned by policy makers in state government at the time, both the demonstration program and the separate R&D program were intended to reap huge economic benefits for Kentucky. National energy policy in the mid–1970s indicated that a coal-conversion industry was going to be established. It appeared that Kentucky, already the leading coal-producing state, needed only to put forth a tangible interest to take maximum advantage of a marvelous opportunity. Eventually an investment in synthetic fuels was made that was unique in its size among the states. Yet, through no fault of its own, Kentucky's program has been only partially successful.

The Kentucky experience raises questions about how far states should go in supporting a national policy, especially in its early stages of implementation. Is it really a national policy, or only national politics subject to change in the next administration? How strong is the commitment emanating from Washington and will it endure? What does the historical record indicate about the factors leading up to that policy? Is the energy shortage really or merely symptomatic of a greater economic malaise? To any state government considering financial participation in a federally conceived program, a word of caution seems warranted: "Be very careful if you ever sense that you are more enthusiastic than the feds are!"

REFERENCES

"Energy: A Special Report in the Public Interest." February 1981. *National Geographic*.

Energy Information Administration. September 1985. "Monthly Energy Review." DOEIA–0035 (85/09).

Holmes, David. May 1983a. "A Commercial-Scale Liquefaction Plant's Impact on Kentucky's Economy." Lexington, Ky.: Kentucky Energy Cabinet.

———. "H-Coal Pilot Plant: Economic Impacts on Kentucky." May 1983b. Lexington, Ky.: Kentucky Energy Cabinet.

"Interview with Edward Noble." February 14, 1986. *Synfuels Newsletter*, p. 506.

Kentucky Energy Cabinet. March 1981. "The Kentucky Synfuel Industry: A Basis for Assessment and Planning." Lexington, Ky.: Kentucky Energy Cabinet.

———. January 1986. "Kentucky Energy Development and Demonstration Program, Summary Report, 1974–1985." Lexington, Ky.: Kentucky Energy Cabinet.

Kentucky Revised Statutes. 1974. Frankfort, Ky.

Stobaugh, Robert, and Yergin, Daniel, eds. 1979. *Energy Future: Report of the Energy Project of the Harvard Business School*. New York: Random House.

6

Class-based Environmentalism in a Small Town: ERDA's "Gasifiers in Industry" Program and the Georgetown, Kentucky, Controversy

ERNEST J. YANARELLA AND HERBERT G. REID

As the effects of the energy crisis of the early 1970s rippled throughout the United States, growing interest in the development of synthetic fuels as a means of responding to looming shortfalls in oil and natural supplies took place in higher policy-making circles. As part of this renewal of state support for energy R&D to convert coal into gas and liquid fuels, the newly instituted Energy Research and Development Administration (ERDA) issued a program opportunity notice (PON-FE–4) in late March 1976 designed to stimulate interest in the integration and evaluation of low-BTU coal gasification technology in operational environments. This federal energy program, known as the "Gasifiers in Industry" program for short, elicited a total of thirteen applications. The three ultimately selected for funding were projects in Georgetown, Hitchens, and Pikeville, Kentucky. Despite a deepening commitment by the Carter administration to an accelerated synfuels development in the late seventies, all three of these projects failed.

This essay presents a case study of the controversy surrounding the rise of public opposition—organized under the banner, "Concerned Citizens of Scott County"—to a joint federal-state decision to construct a miltimillion-dollar, low-BTU coal gasification plant and industrial park in Georgetown, Kentucky. Through a careful analysis of the public record and other available documents and a reconstruction of the dispute through a series of in-depth interviews of principals in the public debate, an effort will be made to determine the factors that precipitated the growth of local opposition to this project and eventually led to its success in achieving withdrawal by the industrial partner in this venture. The failure of this project proved to be a harbinger of the failure of the federal government's effort to commercialize coal gasification technology. Interestingly, it may also anticipate the bases for local opposition within the Georgetown community to the even more massive changes and dislocations that will result from the recent decision of Toyota to locate an $800 million auto-construction plant there.

The following questions are explored: What special resources did this oppositional group possess for mobilizing local citizens against the project? What tactics and strategies were adopted to maximize negative public attitudes toward the proposed gasification plant? How significant was the role played by citizen participation opportunities structured into the decision-making process by environmental statutes and administrative rules? To what extent did debate over technical issues really serve as a smokescreen for issues relating to socioeconomic impact and the "quality of life"? What were the overriding strategies and what were the distinctive strengths, resources, and concerns of the major parties in the controversy (the Energy Research and Development Administration, the Kentucky Center for Energy Research, Irvin Industrial Development, Inc., and the Concerned Citizens of Scott County)? What enduring impact on public perceptions of high-technology energy projects has this controversy had among the Georgetown citizenry? What suitable lessons may be drawn from this experience?

CHRONOLOGY OF THE GEORGETOWN GASIFICATION CONTROVERSY

The environmental dispute over the location and construction of a coal gasification plant and industrial park on the outskirts of Georgetown, Kentucky, began quietly and behind the scenes during the latter part of 1975 and the first few months of 1976. During this period, the industrial developer—Irvin Industrial Development, Inc.—laid plans for the building of an industrial park, first in Richmond and then in Georgetown. Apparently prompted by the desire of Garvice Kincaid, a wealthy and influential Lexington financier, to develop the Georgetown property acquired in 1974 and rezoned for light industry, Irvin Industries began in earnest exploring ways to attract business to the Scott County area. When ERDA issued its program opportunity notice on March 30, 1976, calling for proposals for integrating coal gasification technology into commercial projects, Irvin executives believed they had found a way of selling their business venture to potential industrial-site tenants and obtaining significant public subsidy for the park's energy supply. During the next two months Irvin Industries worked in collaboration with the Kentucky Center for Energy Research (KCER) and a coal committee of the Institute for Mining and Minerals Research (IMMR) on the application for federal and state funding of the proposed project. To enhance the project's attractiveness and perhaps to advance scientific understanding of the low-BTU gasification technology under commercial conditions, Irvin agreed to allow IMMR to secure technical data by monitoring one of the two gasifiers for the first three years of the project's operation. At the beginning of October 1976 ERDA announced the selection of the Irvin proposal as one of six projects that it would fund under the "Gasifiers in

Industry" program, and shortly thereafter KCER communicated to Irvin officials the decision of the state's Energy Research and Demonstration Board to support the construction of the two Wellman-Galusha gasifiers with matching funds from the commonwealth's trust fund.

Local opposition to the project surfaced within a week of the ERDA announcement, emerging in expressions of concern over the economic and environmental impact of the project at the October 8 meeting of the Georgetown City Council. There, in reaction to claims by an Irvin spokesman that the $5.5 million project would generate as many as 6,500 jobs and provide an alternative fuel source for two of the city's largest factories, area residents emphasized the need for controlled industrial growth in the community and raised questions about the possible consequences of the state-federal project for Georgetown's main water source, Royal Spring, and the city's quality of life. Then, at the next council meeting two weeks later, a petition signed by 137 local citizens was brought before Georgetown officials by a representative of Concerned Citizens of Scott County expressing their desire for a thorough evaluation of the project in light of its potential impact upon the local economy and environment. Despite the emergence of local resistance to the project, the City Council signed a conditional "letter of intent" to Irvin Industries indicating its support for the project while calling for its review by the Bluegrass Area Development District.

Largely due to the eruption of local environmental opposition to the project, ERDA asked the industrial developer in a letter dated December 10, 1976, to finance the preparation of an environmental assessment of its project as one of three conditions that would have to be satisfied before ERDA would enter into negotiations for releasing federal funds for the project. Irvin then commissioned an engineering firm, Mason & Hanger-Silas Mason, Inc., to undertake the assessment. In mid–January 1977 the study was completed and released for public inspection. At the urging of the concerned citizens group and with the concurrence of Irvin Industries, the City Council set aside the evening of January 27, 1977, for a public hearing on the environmental assessment. There, amidst one of Kentucky's coldest winters and under the spectre of imminent natural gas curtailments to industries and public buildings, a packed audience at one of the local schools heard the project defended by industrial representatives and state energy policy makers and criticized by Georgetown residents and members of the citizen action group.

Among the most important issues raised by critics against the project were the danger of the contamination of Royal Spring and the degradation of the city's water supply, the capacity of Georgetown's already-depleted skilled labor pool to absorb the anticipated job openings from the plant (now scaled down to 1,600), the ability of plant and park developers to handle safely and properly dispose of ash, oils and tars, and toxic chemical wastes emanating from the plant and industrial processes, and the secondary socio-economic impacts of increased commercial and residential developments upon public services and the quality of life in Georgetown and Scott County. Rejecting the adequacy of the twenty-page engineering report, the Concerned Citizens' legal representative once again called for a complete environmental impact statement (EIS) as required by the National Environmental Policy Act (NEPA) of 1969. Influenced by the persistent calls from the environmental group and its supporters within the community for such a document, the Georgetown City Council approved a revised letter of intent on February 4, 1977, calling upon Irvin Industries to fulfill the NEPA requirements for a comprehensive EIS.

While Irvin Industrial Development, Inc., and ERDA negotiated the specific terms and requirements for producing such an environmental study, the Concerned Citizens of Scott County explored two possible avenues for strengthening its political hand in the dispute. On February 16, 1977, the citizens organization held a meeting under the heading, "Three Alternatives to Tragedy: Thoroughbreds, Tobacco, and Tourism," at William and Jane Allen Offutt's residence that attracted representatives from large horse farms and environmental, historic preservationist, land trust, and horse-racing and other recreational groups. The purpose was to explore the possibility of forming a multicounty umbrella organization to promote responsible land-use policy and controlled economic growth strategies throughout the Bluegrass region. On March 15 the spearhead and de facto leader of the group, Jane Allen Offutt, announced her candidacy for the City Council, thus signaling the group's interest in trying to rebuff the proposed gasification project within the city's major governmental body. With her election to office on May 24 and the victory of at least two other candidates sympathetic to the group's position, the power within the council blunted Georgetown Mayor Warren Powers' behind-the-scenes support for the project.

Throughout much of the twelve-month period after the election, the Georgetown coal gasification controversy was caught up in a lengthy environmental impact statement process. While working with ERDA's choice, the Mitre Corporation, in the preparation of the EIS on the industrial park, Irvin Industries commissioned a Lexington-based firm, Dames and Moore, to perform an environmental report on the impacts of the gasification plant. Then, in part due to the continuing pressure from the concerned citizens group, the U.S. Department of Energy (the administrative successor to ERDA) requested in December 1977 a site-specific water assessment of the project from the federal Water Resources Council. Meanwhile, the local environmental forces pressed their case within the city's political structure, as Jane Allen Offutt gained access to politically strategic positions on the local board of adjustments while her spouse vied for a slot on the city's water board. At the same time, the industrial developers hoped that the release of the Dames and Moore study, finished in September 1977, would be sufficient to obviate the need for the completion and release of the full EIS before the project could go forward.

Public wrangling over the gasification dispute resurfaced in late spring 1978, when the Water Resources Council report and the draft EIS, issued in mid-April and early May, respectively, were subjected to close scrutiny and conflicting interpretation. Critics of the project pointed to the water assessment report's claim that local water sources would likely be inadequate to accommodate the industrial park and future city water demands in time of draught, while supporters noted that the report found little likelihood of any significant impact flowing from the project on the quality of Georgetown's water. The varying responses toward the DEIS on the environmental and socioeconomic impacts of the industrial park were even more contentious. Once more, the U.S. Department of Energy set for August 22, 1978, an open hearing to allow citizens the opportunity to comment publicly on the DEIS and to identify critical and unresolved issues.

At the hearing, held at the chapel on the Georgetown College campus, members of the citizens action group were joined by other citizens representing themselves, the school system, area physicians, the state's Department of Fish and Wildlife Resources, and preservationist groups in opposing the gasification plant and the industrial park. Although federal and state energy officials responded directly and candidly to the ambiguities and uncertainties of the environmental analysis, they also had to confront an expanded list of issues and concerns brought to the public forum by the project's critics: fundamental issues about the fairness and democratic character of the environmental assessment process, the neglect of treatment of the transport of coal and chemical substances and waste products into and out of the park, the future shortage of school space, the threat to water quality and aquatic life in the north fork of Elkhorn Creek, and the aesthetic impact on the historic dwelling and sites located under the shadow of the gasification facility and park.

The denouement of the controversy was characterized by efforts of the local opposition to hold a public referendum on the issue by placing it on the November 1978 ballot. Despite support from the City Council and the Scott County Fiscal Court, the state attorney general's office issued an opinion on September 31 rejecting any legal basis for such an action. Then, State Senator Larry Hopkins, a close friend of the Offutts and Republican candidate for the U.S. House of Representatives, revealed his opposition to the project. Meanwhile, as the Department of Energy moved slowly toward the preparation of the final EIS and a decision on whether to release startup funds for the energy park, its financial backers sought to refute continuing criticisms and negative publicity of the project. With the easing of the natural gas crunch and the shrinking of the federal budget in the context of worsening inflation, federal enthusiasm for coal-conversion technology evidently began to subside. Then, as the concerned citizen forces began to gear up for a possible lengthy and expensive legal battle over the project's development, Irvin Industries sent a letter on February 28, 1979, to Secretary of Energy James Schlesinger asking that its coal gasification proposal be withdrawn.

KEY PARTICIPANTS: THEIR STRATEGIES, OBJECTIVES, RESOURCES, AND LIABILITIES

The Georgetown coal gasification controversy presents itself as a rich, complex, and multidimensional energy/environmental dispute with idiosyncratic and more general features. Insofar as it provides insight into the broader contours of energy and environmental policy making in the United States, it offers some instructive lessons for the course taken in the third round of public mobilization for synfuels development in twentieth-century America. As historical accounts of public interest in deriving synthetic fuels from coal-conversion processes show (Vietor, 1980; Horwitch and Vietor, 1980), efforts at the federal level to trigger heightened energy R&D on this problem were first undertaken in the 1920s against the background of apparent depletion of oil reserves in the Middle East and again in the 1950s for national security reasons during the height of the cold war. Thus, it is not surprising that as a consequence of the threat of oil supply interruptions and the spectre of the "strangulation of the West" by OPEC, the federal government would respond by turning once more to coal conversion as a method of producing alternative gas and liquid fuels.

In order to appreciate the unique and general factors bearing upon the eruption, development, and resolution of this controversy, it is necessary to understand the strategies, objectives, resources, and liabilities that the key participants brought to this political embroglio. While many other individuals, groups, and agencies participated in the dispute, this analysis will restrict itself to the involvement and perspectives of ERDA/DOE, the Kentucky Center for Energy Research, Irvin Industrial Development, Inc., and the Concerned Citizens of Scott County. This analysis will illuminate the political and other factors that brought victory to the opposition forces against seemingly overwhelming odds and provided an early bellwether of the demise of the "Gasifiers in Industry" program.

ERDA/DOE

The institutional origins of the Energy Research and Development Administration may be traced to the growing dissatisfaction with the fragmented character of energy R&D and policy making and to the predominance of support for the development of nuclear technology over all other forms of energy technology in the federal budget since the immediate postwar period (Yanarella and Ihara, 1978, pp. 187–207; Teich, 1977). Until the creation of

ERDA, federal energy activities were dispersed throughout a number of agencies whose primary thrust oftentimes was not directly related to energy problems, or that were narrowly mission-oriented to the development of a single energy source (for example, the Atomic Energy Commission and the Office of Coal Research). Moreover, because of such factors as its unique relationships with the Joint Committee on Atomic Energy, the electric utility industry, and the defense establishment, the AEC was able to command by 1973 upwards of 75 percent of the total energy R&D funding for nuclear programs (Congressional Research Service, 1976, p. 29). To bring an end to these shortcomings, ERDA was created in 1974 through the consolidation of the major energy-related research and development programs and functions previously held by the AEC, the Department of the Interior and the Office of Coal Research, the National Science Foundation, and the Environmental Protection Agency.

The centralization of federal energy activities under one bureaucratic organization did not, however, overcome a number of the deficiencies stemming from its origins. For one thing, the nuclear bias within AEC was largely perpetuated in ERDA due to the fact that fully 85 percent of its first year's funds were derived from the AEC budget and 81 percent of its more than 7,400 permanent positions were filled by former AEC personnel (Joint Committee on Atomic Energy, 1975, p. 11). Moreover, even when the ERDA program turned toward a broader planning vision beyond nuclear power, its technological assumptions and budgetary priorities remained—as noted by the Office of Technology Assessment critique of the ERDA plan and program—skewed "toward a high technologies, capital intensive energy supply alternative" (U.S. Congress, Office of Technology Assessment, 1977, p. 29). Consistent with this emphasis, the ERDA program gave scant attention to near-term energy problems and appeared to emphasize instead long-term options, like breeder reactors, solar-electric systems, and fusion reactors (ibid., p. 39).

In certain respects, ERDA's "Gasifiers in Industry" program may be seen as a weak bureaucratic effort to respond to these criticisms. Although the emphasis in this research program was still on a capital-intensive, high-technology energy alternative, ERDA's program opportunity notice (PON–FE–4) underlined the integration of state-of-the-art gasification technology into "a real industrial application or an operational environment" to meet a near-term energy need (U.S. ERDA, April 13, 1976, p. 9). Its basic objectives were threefold: to demonstrate the technical feasibility and economic potential of low-BTU gasification technology already on the shelf; to amass and analyze technical data from such technology to assist in the development and commercialization of advanced coal gasification systems; and to reduce the consumption of natural gas and fuel oil in the short run until second-generation gasification systems became available (ibid., p. 10). Although many varieties of gasifiers had been prevalent in the United States and elsewhere 30 or 40 years earlier, the program managers added one major stipulation to the type of coal gasifiers that could be proposed by applicants—that they operate under the constraints of prevailing environmental laws and regulations.

This proviso was to play a significant role in shaping ERDA's strategy throughout the local controversy and in enhancing the political position of the citizens action group. With regard to the Irvin proposal, that approach dictated that, once environmental opposition arose in Georgetown, the new federal energy administration would not move to enter into negotiations with the industrial developer until environmental issues surrounding the project were satisfactorily addressed. Thus, as long as old environmental issues remained unsettled or new ones were not confronted, ERDA was willing to allow the project to remain stalled. Even in ERDA's letter notifying Irvin of the conditional selection of its proposal for funding, it set forth three conditions that would have to be met before formal contractual negotiations could begin: (1) obtaining matching funds from the State of Kentucky to maintain a 50/50 cost sharing of the program; (2) receiving commitments from potential users of the coal gas for at least 50 percent of the gas produced by the proposed plant; and (3) documenting assurance that the plan posed no environmental or zoning problems for the local government or community (U.S. ERDA, October 12, 1976).

From the earliest phase of the controversy and at each juncture in the debate through submission of the DEIS, ERDA sought to involve the concerned citizens in meetings and conferences on the Irvin gasification plans or to call for additional, more expensive, and more time-consuming environmental assessments to placate them. After the City Council agreed to sign the conditional letter of intent in early November 1976, ERDA met with representatives of Georgetown, Irvin Industries, the Kentucky Center for Energy Research, and the Concerned Citizens of Scott County to determine whether a more complete environmental study was necessary (Traynor, December 6, 1976). Then, when the public criticism of the Mason & Hanger environmental assessment was aired at the hearing on January 27, 1977, energy agency officials reiterated to Irvin executives that continuing local opposition demonstrated that the last of the three conditions in its letter of October 12 that had to be met for negotiations on the proposed contract to begin remained unsatisfied (Capello, March 18, 1977). As a consequence, a complete and formal environmental impact statement would have to be prepared—thus postponing possible startup of the project by at least another 12–16 months.

The question which naturally arises is: Why did ERDA remain so acutely sensitive to the possible environmental issues and act so deferentially to the citizens action group's concerns? Clearly, the politico-legal setting influencing the nascent synfuels development program in the mid-seventies was dramatically different from the political and legal environment shaping commercial nuclear power development in the fifties and sixties, which had allowed the Atomic Energy Commission (ERDA's predecessor) to proceed without significant political opposition or legal blocks. From 1969 onward, a host of federal laws were enacted—including the National Environmental Policy Act (1969), the Clean Air Act (1970), the Federal Water Pollution Control Act (1972), the Toxic Substances Control Act (1976), and the Clean Air Act Amendments (1977). Each of these laws came to impose federal constraints or mandate regulatory standards upon environmental pollutants emanating from coal use and coal-related technologies. No less important, as Walter Rosenbaum has noted (1981, p. 186), "These measures provide the legal standing and the substantive grounds for citizen groups and government regulators to apply legal restraint on synfuels technologies should they threaten to become ecologically dangerous." In short, ERDA's policies and programs were now operating in an institutional climate heavily affected by the enactments of the environmental decade (Anderson, 1973; Wenner, 1976, 1982).

If these laws and the accompanying mission definitions of energy agencies like ERDA broadened public and legislative access to regulatory questions and required more formalized opportunities for public participation in energy/environmental decisions and programs (Bishop, 1975; Gellhorn, 1975), the crisis atmosphere of America's energy woes in the decade of the seventies also triggered opposing administrative urges to streamline siting and permitting processes and to bypass public review of various energy activities posing possible hazards to the environment and public health. The Carter National Energy Plan II, for example, included the proposal to create an energy mobilization board to speed up construction of energy-production projects deemed vital to American energy independence and national security (Goodwin, 1981). In the case of the Georgetown controversy, the letter written on March 18, 1977, to Irvin vice president Tom Lohr by ERDA assistant director of operations Joseph Cappello revealed the conflicting political-democratic and authoritarian-bureaucratic impulses operating within the federal energy agency.

In that letter Cappello indicated that, on the one hand, ERDA had "become convinced that Irvin will be unable to demonstrate compliance with Condition #3 so long as local opposition to the project continues and the City declines to grant unconditional approval." On the other hand, Cappello assured the corporate vice president, "Fortunately, the National Environmental Policy Act of 1969 (NEPA) provides ERDA with the mechanism to determine *for ourselves* whether the project poses no 'environmental problems to the local government or community.' " The instrument was a detailed environmental impact statement that, Cappello indicated, "ERDA is prepared to undertake with Irvin's cooperation and assistance." In effect, the assistant director signaled the industrial developer that the formal preparation of this statement would allow ERDA independently of continuing environmental opposition or city council refusal to grant unconditional approval "to conclude fairly whether Condition #3 is met and to fulfill its obligations under NEPA." When this communication was eventually discovered by the concerned citizens leader, Jane Allen Offutt, she raised a furor at the public hearing on the DEIS over its antidemocratic thrust and intent, much to the embarrassment of the U.S. DOE representatives (Offutt testimony, in Proceedings of a Public Hearing, August 22, 1978, pp. 25–27).

Whether this authoritarian temptation would have overriden the formal democratic procedures instituted by NEPA and the other laws and regulations promulgated during the high tide of the environmental movement became a moot issue when Irvin decided to pull out of the venture before the release of the final environmental impact statement and the Energy Department's decision to begin contractual talks with Irvin. Despite initial enthusiasm for low-BTU coal gasification technology within the federal energy administration, by 1979 the sense of urgency animating national energy R&D had diminished considerably due to new DOE estimates of the abundance of natural gas and severe cutbacks in the DOE budget for fiscal 1980 occasioned by galloping inflation ("Changes in U.S. Energy Plans," March 22, 1979). When the fiscal year 1980 energy budget cut more than $1 billion in funds for conversion plants, it became expedient for the Department of Energy to maintain the stance that "until [environmental] questions are resolved we aren't going to proceed with it" ("Damper is Placed on Kentucky's Hopes," January 23, 1979, p. 7).

Kentucky Center for Energy Research

If the Arab oil embargo and the accompanying "energy crisis" gave impetus to centralization and coordination of administration over federal energy programs, these events also prompted many states in the nation to undertake similar administrative reorganization either to take advantage of indigenous energy reserves or to foster serious conservation measures to reduce energy demand in the face of rapidly escalating energy prices. Under the leadership of Governor Wendell Ford, the Commonwealth of Kentucky initiated a series of organizational innovations in state government to find ways of utilizing its estimated 65 billion tons of coal reserves for the state's economic development and as a means of responding to calls for laying the foundations of America's energy security in the future.

Beginning in 1972 with the initiation of a research program on coal gasification at the Institute for Mining and Minerals Research, the state moved in 1974 to build a public infrastructure for advanced means of coal utilization by increasing energy R&D to $3.7 million and by establishing a $50 million energy-development trust fund for state participation in pilot and demonstration projects for the conversion of coal to gaseous and liquid fuels. It was from this trust fund that the three low-BTU gasification projects were to be supported. In August 1975 a Department of Energy within the Kentucky Development Cabinet and the Kentucky Center for Energy Research were created. The Kentucky Department of Energy was vested with the tasks of managing the commonwealth's energy programs (including conservation, allocation, and resource-development planning), while KCER was assigned the responsibility for administering the state's energy research and development activities through the Energy Development and Demonstration Trust Fund and the state's general funds.

The state, partly taking the lead in promoting the promise of coal conversion technology, endeavored to assume "the leadership in the formulation of a *comprehensive statement* of what national policy should be with respect to the future use of coal as the nation's major energy resource and the development of the coal industry" (KCER, June 1976, p. 13). In succeeding years, however, this aggressive strategy for exploiting the state's coal resources through advancing technology ran into imposing hurdles. From a review of the administrative strategy and roles that KCER adopted in the Georgetown/Irvin controversy, it is evident that perhaps its most serious problems lay in the unresolved tension that had developed at the state and federal level between politics and administration. Moreover, this state energy research agency charged with energy policy planning and with technological administration for realizing larger public policy goals could not plan or administer due to the imperatives of the surrounding political system and the constraints of the larger political economy of coal in Kentucky.

The problem of an extensive state planning apparatus that cannot plan is not unique to the commonwealth; indeed, it is a problem endemic to American policy-making agencies generally in the technocorporate phase of our political economy (Reid and Yanarella, 1980; Wolfe, 1977; Burton, 1980; Daneke, 1980). As the United States emerged as an industrial and world power with extensive global commitments in the postwar era, the conflicts between private power and the goals of public policy were largely resolved through state accommodation to the realities of corporate power and elite consensus. Against this historical background, the function of state administrative agencies was to maintain optimal conditions for capital accumulation, in part by socializing private costs and risks (in investment, technology, and the like) through direct public subsidies in such areas as R&D and indirectly through tax incentives. Administration at federal and state levels assumed a management role relative to the supervision of private activities supported by public investments and the maintenance of optimal business conditions for economic development (Wolfe, 1977). But in comparison to France, where the state takes a more directive role in national planning and where a *dirigiste* tradition exists and legitimates this state function even in the business community, state and federal governments in the United States have adopted a willing, but weak, supervisory role and have been reluctant planners—in part because of the continuing hold of the myth of the market in American political culture and in part because of the penetration of corporate power into formal political institutions.

The interpenetration of political and economic power and the extreme weakness of the supervisory function of Kentucky's administration over energy R&D were everywhere in evidence in the Georgetown dispute. In this controversy, the overall strategy of the Kentucky Energy Center was to support, subsidize, and collaborate with industry in an effort to realize the construction of the gasification plant and industrial park while simultaneously mollifying public concerns about the environmental and socioeconomic impacts upon the community. Its own objective, however, was a technological one: to facilitate the building of a demonstration coal gasification plant that would exhibit the feasibility of producing gas from coal and to foster further development of more advanced technological designs. Yet its administrative framework was permeated by corporate influence and priorities.

An offspring of the state's development cabinet, KCER was required by executive order to bring any proposals for support by the Energy Development and Demonstration Trust Fund before a fourteen-member Board of Energy Research heavily dominated by coal industry executives who could wield important power over the choice of energy R&D programs and policies vis-à-vis engineering administrators and political officials who also sat on the board. In addition, despite potential conflict of interest, fluid traffic of interested parties between state energy administration and the corporate world or local government took place in at least two cases. A former vice president of Industrial Develoment, Inc. (IDI), the predecessor of Irvin Industries, was chosen as the KCER administrator, and a legal expert on the KCER staff was elected Georgetown city attorney.

The dual role played by the IDI executive cum state energy administrator in the project was significant. For example, during his tenure with the private development firm he had a direct hand in the original decision in 1975 to select Georgetown over Richmond as a site for the gasification facility and industrial park. His instrumental role in influencing the site choice went far in answering the oft-asked question among Scott County opponents: why Georgetown and not somewhere else? Then, when he became the KCER chief, he stood at the helm guiding the state's potential involvement in the project from the submission of the cost-sharing request for the construction of the gasification plant to the Kentucky Energy Research Board and the attempt to obtain a state-built access road to the industrial park property through the other tasks and phases of state partnership in the venture.

Given the tightness of interface between the state energy agency and industry, KCER had very little political leverage over the industrial developer in their dealings on the project even though the KCER administrator and staff periodically expressed chagrin toward various Irvin Industry promotional activities during the controversy. Early in the dispute, when Irvin was called upon to prepare an environmental assessment, internal memoranda within KCER reveal intense displeasure with Irvin's choice of the engineering firm that had performed earlier design work on the gasification plant proposal as the consultant to undertake the environmental report. The memoranda also reveal KCER's harsh criticism of the shoddy, inadequate character of the twenty-page document in addressing the concerns of the environmental opposition (KCER staff memorandum, December 26, 1976; Traynor, January 13, 1977). Later, as Irvin was preparing a fuel gas brochure highlighting the nature of its project and its technological and economic potential, the KCER administrator and staff member were so discouraged by its heavy-handed promotionalism and its virtual technical illiteracy of the gasification process that the latter wrote: "I fully concur in your comments on the Irvin draft brochure. If Irvin put this out you would be obliged—in the public interest and to protect other low BTU gasifier supporters—to dissociate KCER from it" (Traynor, October 10, 1977). Yet in these and other cases of dissatisfaction with the industrial partner, the KCER staff dutifully offered technical assistance to "save" Irvin from its worst faults and errors.

The Kentucky Center for Energy Research, lacking an institutional foundation for asserting a strong supervisory role and caught in a political situation that undercut its planning function, could at best play a supporting role to requests and initiatives emanating from the industrial developer. This general role translated into four specific functions taken on by KCER in the Georgetown project. First, the state energy research agency drew upon the development trust fund to commit matching funds ($2.776 million) with ERDA to pay for the construction of the gasification plant (Kentucky Energy Research Board Minutes, October 19, 1976). Second, since ERDA would not agree to provide Irvin with federal support until Condition #3 was met, KCER agreed to underwrite one-half of the cost ($29,000) of the environmental impact statement (Drake, June 3, 1977, and July 27, 1977).

Third, throughout the project dispute KCER secured or generated important technical data and provided extensive staff support for Irvin Industries—all of which were crucial to increasing the prospects of Irvin's success in winning funding for the energy park project. For instance, KCER derived far more reasonable cost estimates of the coal gas than those originally quoted by Irvin Industrial officials (McGurl comments, in Proceedings of a Public Hearing, August 22, 1978, p. 104). Similarly, its staff generated a lower and more realistic estimate of the number of jobs that would be created by the plant and park than the employment estimates that quickly got Irvin promoters in trouble with the community (Sanders remarks, in Proceedings of a Pubic Hearing, January 27, 1977, pp. 14, 15, 17, and 21). KCER staff members, furthermore, solicited data on coal usage of other operating low-BTU gasification plants, drew up bid invitations for EIS preparation for Irvin to send out to environmental and engineering consulting firms, reviewed and commented on the original draft of the Irvin funding proposal to ERDA, wrote background reports to serve as Irvin publications, and drew up environmental assessment and EIS schedules for the industry to follow.

Finally, KCER acted as a bridge between industry and the state and federal government and between industry and the local community. In its partnership role, the energy research agency was continually called upon to promote facets of the project (like the access road) with other state departments, monitor for Irvin Industries key developments taking place in ERDA, and assuage irate letters from concerned citizens. A key problem in acting as a subservient instrument of corporate development programs is that those functions designed to promote industry's capital accumulation and the commonwealth's economic development oftentimes come into conflict with the other major role of state and federal organizations—the fostering of legitimacy or public credibility in the representative character of governmental institutions. In the case of KCER, one of its important legitimacy functions is to assure citizens of the openness of administrative processes in energy policy-making and of the agency's concern to foster environmental protection and public health within the context of advancing technological development in the energy realm. It is not surprising that one negative consequence of KCER's supportive role of Irvin Industries is that public criticism of the joint venture was deflected onto the state (both the federal and state agencies), thus further eroding the bases of citizen support for state administration and planning (see Jane Allen Offutt's comments in Proceedings of a Public Hearing, August 22, 1978, pp. 25–27).

It is small wonder that KCER's director of the demonstration projects, in the light of the failure of the commonwealth's synfuels program to date, should deny the possibility of comprehensive planning for energy development and instead should rationalize ad hoc state support for corporate initiatives and opportunities as they arise (Mitchell interview, June 1, 1982). Yet this was precisely the strategy that simultaneously undercut the political bases of support for the energy agency's planning function and contributed to the synfuel program's troubles. His major conclusion drawn from Kentucky's experience in the Georgetown case, however, does seem appropriate: KCER should "never again . . . enter into a project solely with another government body" (p. 22).

Here, the chief beneficiary, he noted, was going to be a company with no substantial monetary or other commitment invested in the project (p. 23). From the very beginning Irvin chose to minimize its financial obligations and risks while allowing state and federal energy agencies to shoulder the full cost burden of the gasification plant. A state energy agency with strong supervisory powers and a de facto planning capacity might have rejected this project proposal or at least negotiated a fairer division of risk betweeen public institutions and industry. Quite simply, the Kentucky Center for Energy Research lacked those requisites. Unfortunately, in the political environment that exists today, the conditions for reshaping energy organization in the commonwealth to meet the challenge of our energy future in a way that gives Kentucky coal a legitimate role do not yet appear even on the distant horizon.

Irvin Industrial Development, Inc.

When Irvin Industrial Development, Inc., began its promotional campaign for the gasification plant and industrial park, it was not exactly a household word in Georgetown, much less in the Bluegrass. Indeed, the corporation did not exist prior to early 1976, when it was formed through a real-estate deal between a powerful Kentucky banking interest and an industrial developer. Its origins lay in the purchase in 1973 by Danville Realty of the 172-acre plot of land located just south of Lemons Mill Road in Georgetown and the efforts by the realty company to develop the land for commercial purposes. Shortly thereafter, Danville Realty was tranformed into an industrial development firm called Industrial Development, Inc., the development arm of its parent company, the Kentucky Group Banks, part of the Kincaid financial empire. IDI originally considered developing into an industrial park a parcel of land located in Richmond, Kentucky. But because of the need for greater financial backing to initiate the project and the larger size of the Georgetown property, it shifted its plans to Scott County. When ERDA issued an early Request for Proposals (RFP) to solicit interest in industrial proposals for federal support of the construction of low-BTU gasification technology for industrial uses, Industrial Development, Inc., believed it had found a means of receiving public subsidy for an alternate fuel source to natural gas (whose use was being curtailed due to shortfalls) that would make its industrial park idea attractive to potential industrial occupants. As a vehicle for fostering these plans at the Georgetown site, Irvin Industrial Development, Inc., was formed in 1976 by the merging of Industrial Development, Inc., and Irvin Industries, Inc., a Connecticut- and Kentucky-based multinational corporation.

The chilly reception of its energy park proposal from the Georgetown community stemmed from a number of factors. In the first place, the image of Lexington financier Garvice Kincaid as an aggressive, powerful banking mogul and corporate wheeler-dealer did not sit well with many Georgetown citizens. When Kentucky Group Banks bought First National Bank of Georgetown, many residents of this staid, conservative town reacted negatively to its banking innovations and shuddered at its possible impact upon the future course of Georgetown's economic development. Also, a few years before the environmental dispute Industrial Development, Inc., had stirred up controversy and antagonized some in the community when it began to construct a building on the Lemons Mill Road property without obtaining the requisite permits from the local planning and zoning board (Sutton remarks, in Proceedings of a Public Hearing, August 22, 1978, p. 60; Snyder interview, July 12, 1982, p. 11). Although Kincaid himself died before the project received any publicity in the community, the legacy of his aggressive business style and past dealings with the Georgetown community was to dog his successors, Al Florence, Thomas Lohr, and Kenneth Holbrook.

Irvin Industry officials, totally insensitive to the controlled-growth philosophy guiding the views and perceptions of an influential segment of the local community and oblivious to its zoning and planning tradition, embarked upon a strategy to win approval for its energy park scheme that grew out of the aggressive development approach identified with the Kincaid Kentucky Group Bank empire. Illustrative of its heavy promotion orientation was its campaign, begun in early April 1976, to interest prospective tenants of the industrial park, dubbed the Kentucky-Federal Energy Park, even before federal support or state commitment was forthcoming (Lohr, April 20, 1976). It also revealed its aggressive approach in its unsuccessful effort to persuade ERDA and KCER to accept a "fast-track" method of construction of the plant that would have allowed it to run roughshod over environmental assessment requirements and perhaps zoning ordinances (Lohr, October 6, 1976; Kentucky Energy Research Board, Minutes, October 19, 1976).

But the most glaring evidence of its promotional/developer mentality was the approach it adopted in presenting its proposal to the Georgetown community at the October 6 City Council meeting. This presentation by one of Irvin's vice presidents amounted to a tremendous industrial development sales pitch for a project that would generate upwards of 6,500 new jobs, transform the economic face of Georgetown from a predominantly agriculturally based economy to an industrial economy, and require significant increases in the town's housing and public services, especially schools (Holbrook's comments in "Gasification Plant Development," October 13, 1976, p. 1). Although it mobilized some enthusiasm and support from the Georgetown mayor and the town's newspaper editor, this heavy-handed promotionalism triggered unorganized but influential middle-class professionals and upper-income individuals into action under the banner of "Concerned Citizens of Scott County."

This industrial developer mindset was manifested as well in its lack of any real technical experts on the staff and apparently its lack of a firm conception of the character of the industrial park it wished to develop. As one KCER staff member concluded, much of its promotional material on the coal gasification plant was "similar to what Gen. Motors uses to sell their cars" and nothing more than "an insult to industrial administrators and engineers" (Traynor, October 10, 1977). Moreover, due to its cloudy understanding of the nature of the park and its potential occupants, it immediately got into trouble over its exceedingly high estimates—apparently pulled out of the air—of likely new employment stimulated by the project in a community that was experiencing monthly unemployment fluctuating from 2 to 3 percent. From October 1976 through August 1978 Irvin revised its job estimates from 2,900 to 6,500 to 1,600 to 450–500 to 390. Another area where lack of technical expertise hounded it throughout the debate had to do with whether the energy park would house light industry or heavy industry. Originally, when the land resided in the hands of IDI, it was zoned for light industry. However, because the industrial park could only use efficiently the fuel gas produced by the coal gasification system, most of the industrial users should have logically been characterized as heavy industry by the local planning and zoning board. And, although Irvin officials continued to try to placate the environmental opposition by assuring them that the zoning board would evaluate each industry on a case-by-case basis, they remained inconsistent in their publicity on the issue. Sometimes they suggested to the local community that only light industry would be permitted, while at other times they offered to ERDA and environmental consultants working on the DEIS scenarios of potential park inhabitants that were predominantly heavy industry gas users (Lohr, October 7, 1977).

While pursuing an aggressive promotional strategy on the project with the public and with potential park tenants, Irvin Industrial Development, Inc., engaged in a low-risk/high-payoff approach vis-à-vis its federal and state partners. That is, Irvin strove to displace as much of the burden of cost and risk upon ERDA and KCER as those administrative agencies were willing to shoulder. Continually portraying itself as a small company that could ill afford to expend large sums of venture capital on the energy park (for example, Lohr, June 10, 1976), it sought to elicit "front money" from KCER to begin the feasibility studies for the Georgetown facility (Lohr, April 9, 1976), and requested state and federal subsidies for funding the preparation of the environmental assessment and impact statement (Lohr, December 1, 1976; Drake, June 3, 1977). When the Kentucky Energy Research Board presented it with two options for repaying the state share of the gasification construction costs, it chose the "no capital risk" option that freed it from investing any capital for the construction of the plant and only recompensed the state when the project began to turn a profit for the corporation (Kentucky Energy Research Board, Minutes, October 19, 1976, p. 3; Drake, October 25, 1976, and November 17, 1976).

The great strength of Irvin's overall strategy for promoting the project, however, resided in its capacity to line up political backing for its industrial development venture. In early 1976, when ERDA issued a predecessor to PON-FE-4 (RFP No. E [49-18]-2043) that assumed tighter public control of any patents deriving from advancements in coal gasification technology and required contractors to return the government's funds at fair market value upon completion of the project, Irvin officials unleashed a volley of criticism directed at ERDA and embarked upon a vigorous political campaign enlisting Kentucky's Senator Ford to revise the RFP deemed unacceptable to them. When state commitment to cost sharing of the project with ERDA was delayed, they once again turned to state representatives (Senators Ford and Huddleston and Congressmen Natcher and Perkins) to apply their muscle to key access points in state government to hasten the process (Lohr, June 10, 1976, and October 6, 1976).

The Irvin people thus presented themselves to key participants in the debate in a variety of guises. To the concerned citizens group they were outsiders intent on despoiling the natural landscape and visiting drastic and destabilizing consequences upon the local community and economy. To the state energy policy makers they were technically incompetent but politically well-connected industrial adventurers with public money who were the only game in town. To ERDA bureaucrats they were a resource opportunity to evaluate existing coal gasification technology for later technological advancement if they could overcome local opposition by negotiating the environmental impact process. To environmental and preservationist groups they were aggressive land developers whose project must be made to submit to the letter of NEPA regulations. According to their self-portrait they were a young, budding corporation offering economic growth to a small town and responding to serious energy shortages in a risky business climate requiring public subsidy in order to turn a profit.

Their exit from the federal-state venture on February 28, 1979, brought to an end a lengthy and acrimonious dispute. The events that helped to trigger Irvin's withdrawal also foreshadowed the demise of the "Gasifiers in Industry" program, for they demonstrated how precarious were the conditions that ushered it into existence and how easily those favorable circumstances could be upended.

Concerned Citizens of Scott County

When disgruntled citizens of the Georgetown area began to organize, protest, and petition their local elected officials over the environmental and socioeconomic consequences of the proposed gasification plant and industrial park in their city's outskirts, they were reenacting forms of citizen action and appealing to certain rights and prerogatives of civic republicanism—a time-honored tradition going back to the earliest days of the American republic (Bellah et al., 1985). Yet the outcome of their political efforts was never foreordained, since the political constellation arrayed against them included a growth-oriented mayor and newspaper editor, a powerful financier and his interlocking banking and corporate network, the state energy agency, and the federal energy administration, as well as one of Kentucky's most popular and influential senators.

Not least of all, the Concerned Citizens of Scott County was laden with certain characteristic strengths and weaknesses of the organizational form that this protest took: a single-focus, local citizens action group. As the critical literature discloses (Caldwell, 1976; Boyte, 1980; Berger and Neuhaus, 1977; Cloward and Piven, 1979; Wellstone, 1978), citizen protest organizations are often fragile, short-lived political instruments of change prone to self-dissolution over the long run because of fluctuating, changing memberships, flagging energies related to an inability to institutionalize themselves organizationally and financially, growing tendencies among the leadership core toward elitism or vanguardism in the face of quixotic mass citizen support, limiting strategies of obstructionism based upon a narrow not-in-my-backyard philosophy, and generally overwhelming inequities of political, monetary, informational, and other resources vis-à-vis government institutions and entrenched economic interests (Yanarella, 1985; Humphrey and Buttel, 1982). On the other hand, their weaknesses and limitations are sometimes compensated for by the emergence of creative and talented leadership from within their midsts, the liabilities and rigidities of their more powerful bureaucratic and corporate opponents, the formal procedural rights of public information and participation in administrative and regulatory processes, the beneficence of the American legal system, which periodically decides in favor of the citizens group on a point of law, and the capacity of their underdog status to mobilize within the larger public sympathy and support for their fight (Boyte, 1980).

The following sketch of the citizens action group can be drawn from interviews and news accounts. The Concerned Citizens of Scott County drew disproportionately from the middle and upper middle-class and professional stratum of the Georgetown community and had its ideological base in mainstream to conservative political beliefs (Phares interview, June 14, 1982; Hall interview, June 28, 1982; Snyder interview, July 12, 1982; Graddy interview, July 16, 1982). This class bias was revealed, first of all, in the dominant class composition of its membership. Although the key members chafed at the "landed gentry" label used to characterize the group by the opposition (Phares interview, June 14, 1982, p. 1), a review of the group's most active core and the significant others whom they contacted and successfully mobilized in their campaign to overturn the joint state-federal venture with Irvin Industries reveals a decidedly upper-class orientation. For instance, among its key leaders were a veterinarian with strong connections to horse-farm families, the spouse and daughter of a Georgetown bank executive, a mortgage and real-estate agent with long family ties to the area, a high-powered Lexington lawyer in a prominent law office, an opthalmologist and university professor with a lucrative practice, his spouse, and numerous professionals from Georgetown College and within local and state government—including an air-pollution expert who clandestinely passed on damaging environmental information secured in his office. In addition, the group was able to call upon the essentially gratis services of a high-powered lawyer in a prominent Lexington law office, timely aid and support from two of Georgetown's largest factory employers who were concerned with the impact of the industrial park upon wage scales and competition for skilled labor in an already depleted local labor market, and the environmentalist and preservationist concerns of middle and upper-middle-class members of the Sierra Club and area land

and trust organizations populated largely by Scott County families of wealth and position.

The other area in which their distinctive class bias was expressed lay in the "political" and "impact" elitism (Humphrey and Buttel's term, 1982, p. 132) characterizing their political goals and actions. In public-education flyers and statements before public meetings and hearings on the proposed project, concerned citizen spokespeople continually drew a portrait of Georgetown as a small-town haven, a "garden spot," blessed with a healthy economy, low unemployment, and beautiful, green surroundings populated with horse and agricultural farms. Drawn from the middle-class professional and upper middle-class stratum of Georgetown, these individuals and families could enjoy both the satisfactions of small-town life and the benefits of Lexington while railing against their urban neighbor for its ugliness and city-related problems (pollution, crime, traffic, and haphazard zoning and overdevelopment). In their minds, Lexington presented itself as a negative model of economic growth that they sought to avoid replicating in Georgetown. As the history of the city's zoning and planning efforts discloses, Georgetown has greatly benefited from such factors as the federal impetus to encourage local communities to plan, the planning failures of Lexington, and the tendency of this Bluegrass urban center to throw off to neighboring communities trailer parks and other urban blights it did not want (Snyder interview, July 2, 1982, pp. 1–4).

Still, this image of the Georgetown community and its need for controlled growth must contend with certain demographic realities of the city and the surrounding county. At that time Georgetown was a small town of nearly 9,000 residents while the county commanded a population of 18,654 (U.S. DOE, April 1978, p. S–4). For the city and the county, median family income was $8,193 and $7,568, respectively. But in spite of its much-vaunted low unemployment (ranging from 2.5 to 3.5 percent during the dispute), the Georgetown population contains a sizable lower-income stratum that includes many of its black citizens, who account for 17 to 18 percent of its total populace. Moreover, the city has been dogged by the continuing problem of trying to stem the drain of its young people who have not been able to find employment in the area. In the citizens group's campaign and appeals to the Georgetown citizenry, the plight of youth, working people, and the black minority who were seeking jobs in the area were never seriously considered.

This point becomes even more relevant when one realizes that the political fulcrum of the community, then and now, lies with the lower-income Democrats (Snyder interview, July 12, 1982, p. 10). Upper-income Republicans, like the Offutts and many of their fellow members and allies in the community, must either sacrifice their vote by registering Republican or sign up as Democrats and suppress or camouflage their ideological orientation or party identity. In the case of Jane Allen Offutt, her bid for the City Council succeeded largely because she could draw upon an active Republican base of support in a general, formally nonpartisan election. But her attempt to run for mayor in the Democratic primary foundered due to the exclusion of many of her Republican supporters. This analysis also casts doubt on the outcome of the proposed referendum on the Irvin energy park. Depending upon whether the ballot would have been citywide or countywide, the concerned citizens would have faced a close battle involving the upper-class advantages of an off-year vote and the working-class and lower-middle-class strength of the larger voting constituency.

The primary manifestation of the impact elitism of the Concerned Citizens of Scott County lay in its environmentalist concerns—and particularly in its frequent use of such issues to conceal the class-based concerns for socioeconomic impact and quality of life that motivated most of its membership. Neglecting the legitimacy of the claims of the working class and the black minority in the community, the citizen action group marshalled its talents and resources and set upon a course of political action intended to stretch out the environmental impact process to stall the gasification and industrial park project. At various stages in the controversy, the organization's leaders entertained a number of possible avenues for implementing this political strategy. At first, the group attempted to block the project by putting pressure on the City Council. Then, when it appeared that it might have to turn to the courts for redress, some of its members thought about the vehicle of a multicounty alliance of environmental, preservationist, recreation and tourist, and horse-farm interests. In addition to raising large sums of money for legal fees, this umbrella organization held out the promise of putting in place an areawide Bluegrass coalition to monitor and respond to local and state economic development plans that were seen as inimical to the region's environmental health and to controlled economic growth. This more perceptive, long-haul approach to the problems raised by the controversy, however, was dropped due to its lukewarm reception by the diverse groups and to Jane Allen Offutt's decision to seek election to the City Council. With her election, the political strategy of the organization became set and the long-term impact of their battle largely determined.

More concretely, this political strategy involved an approach to political conflict that dictated the continual effort at each phase in the debate to set the agenda of issues in dispute and, when success was not achieved, to widen the arena of conflict in order to alter the stakes. As E. E. Schattschneider (1960) has argued in his now-classic statement of democratic politics in America, in a pressure system of interest-group politics biased toward the haves over the have-nots, the major chance of increasing the potential for victory of the underdogs is the strategy of expanding the scope of conflict to overcome the advantages of the topdogs. Otherwise, the natural tendency will be for political issues to be settled quietly and behind the scenes by government and business elites. As is evident from the brief chronology of the controversy offered earlier, in the absence of the concerned citizens forces adopting essentially the strategy outlined above, it is very likely that the perfunctory environmental report offered by the Mitre Corporation at the preselection stage of ERDA's review process would have been regarded as sufficient to meet ERDA's institutional obligation to fulfill to NEPA requirements. Instead, the citizens action groups removed the issue from the "iron triangle" of business, political, and regulatory elites and recentered the decision-making process within the arena of citizen politics.

What then were the distinctive strengths residing in the concerned citizens group that helped them succeed in defeating the Irvin gasifier/industrial park plan? Certainly one of its key assets lay in the special mix of talents and personal capabilities of its leadership core. Each of the core members brought unique abilities and knowledge from the real-estate and mortgage world, from the banking and business realm, from the farming and horse-raising sector, and from the academic and scientific domain—all of which strengthened the hand of the group in challenging apparently unbeatable foes. In addition, Concerned Citizens of Scott County was blessed with a hard-driving and indefatigable leader and catalyst who showed consummate political skills in mobilizing a broader circle of friends and contacts to assist in putting pressure on the political center while using her elected position on the City Council to legitimate the issues and perspectives of the group in the eyes of the wider public and to channel protest into local government through her elective post and her selection to other influential positions on governmental bodies that could block or delay further the course of the decision-making process (Phares interview, June 14, 1982, pp. 12–13, 21–22; Snyder interview, July 12, 1982, p. 12).

A third key strength of the local opposition lay in its ability to link its localized protest to national environmental trends skeptical of high-technology projects and their uncertain impact upon the environment and the quality of life—a condition that Allen Mazur (Mazur, 1975, p. 73; Leahy and Mazur, 1980, p. 275) has argued has often been a decisive factor in determining the success or failure of public opposition to such large-scale technological activities. Beyond this, Mazur has also noted that the strength of public and environmental opposition is also frequently tied to increasing mass media coverage, whether positive or negative (Leahy and Mazur, 1980, p. 279). Interestingly, the extensive media coverage of the gasification dispute in the local newspaper and the campaign of its editor to discredit the concerned citizens' views on the project's impact for the community seemed to play into the hands of the local protest group by publicizing their issues and perspectives and giving them greater legitimacy in the eyes of the wider public.

Finally, in its battle the Concerned Citizens of Scott County was able to mobilize support from historical preservationists and conservationist groups in the area by questioning the dangers of the project to Royal Spring (a source of Georgetown's water supply and an historic site to which many representatives of these groups attached great symbolic importance), just as it was successful in splitting the local business community (downtown businesses versus larger industrial firms) by gaining allies among the large industrial firms who feared even keener competition for Georgetown's labor market in the event the industrial park became a reality. From its self-image of being a poorly funded, loosely organized, and politically weak David fighting against a well-financed, politically powerful, and seemingly unbeatable Goliath, the local opposition was successful at each crucial turn in the dispute in using its impressive tangible and intangible political resources to enlist wider sources of public support and to enlarge the arena of political conflict—ultimately outlasting the patience and will of the industrial firm whose project it was protesting.

Concerning the long-term impact of the concerned citizens' efforts and their eventual victory, the record is at best a checkered one. On the one hand, the local protest group did succeed in fortifying and advancing the city/county commitment to planning and zoning into the eighties. Several former members of the group participated on a county industrial development authority to oversee the controlled economic growth of the Georgetown area (Phares interview, June 14, 1982, pp. 14–15; Ashley interview, July 11, 1982, p. 10). In addition, Jane Allen Offutt chaired the citizens committee to rewrite the Comprehensive Plan for Georgetown and Scott County to direct the growth plans of the city and county for the succeeding five years ("Offutt Announces," April 18, 1979, p. 10). And two government officials in Georgetown have credited the concerned citizens organization with having a diffuse influence in shaping public attitudes in favor of local planning and controlled growth (Snyder interview, July 12, 1982, p. 14; Ashley interview, July 11, 1982, p. 11). Yet in the recession of 1980 Georgetown's unemployment rate mushroomed to over 10 percent and left the community vulnerable to the state's decision in late 1985 to woo Toyota Corporation to Scott County to build an $800 million automobile assembly plant—a decision that will likely have enormous social, economic, and environmental consequences for Georgetown and Scott County residents.

While the forces that helped to galvanize local opposition to the earlier project remain entrenched in the community, the absence of any institutionalized organization or forces stemming from the concerned citizens protest has meant that thus far only small business remnants within the Lexington and Georgetown community have sought to mobilize public opinion against the largely state-directed action. Whether a different strategy—one involving the construction of the kind of central Kentucky environmental/controlled-growth alliance that the local group courted early in the dispute—could have compelled the Collins administration to take into account at the outset the environmental and socioeconomic impacts of such a major industrial development is problematic. What is clear, however, is that the consequences of this decision will far outweigh the anticipated impacts flowing from the Irvin industrial development scheme.

CONCLUDING LESSONS OF THE CONTROVERSY

Various lessons for differing parties can be derived from this dispute. Citizen action groups opposing high-technology projects might conclude that a strategy of adamant political and (threatened) legal opposition on environmental grounds, coupled with a sustained effort to widen the parameters of conflict and draw ever-broader elements of the populace into the debate, is likely to be a winning strategy as long as state and federal environmental regulations afford citizens openings for public information and participation. State energy policy makers will probably draw two lessons: (1) careful prior assessment of local conditions (including socioeconomic factors, environmentalist and conservationist attitudes, local labor supply, and the like) is essential before commitment of state technical and financial support should be made for demonstration projects; and (2) risky industrial ventures for the state involving minimal commitments from the industrial partner must not be chanced unless the distribution of potential risk and benefit can be made more equitable. Second, the Georgetown episode would also suggest to energy policy makers that only infrequently can political support generated by state agencies on behalf of a poor, or at least questionable, high-technology industrial development project overcome inherent weaknesses in such projects.

This case teaches social scientists that quality-of-life considerations and the much-touted "environmental ethic" have deeply and widely permeated elements of the public (whatever their ideological predispositions), drawing particularly strongly upon rural attachments to the land and to symbols of a local heritage. As the Georgetown case clearly illustrates, these concerns can serve as potentially mobilizable political resources for environmental and controlled-growth advocates opposing at local levels issues like the siting of energy-generating plants and industrial development centers. Similarly, if there is a lesson to be learned by corporate developers from this experience, it is that this deep-seated and continuing public support for environmental protection and public health should compel corporate officials to treat the NEPA requirements for environmental impact assessment not as a pro forma exercise but as a serious matter.

Finally, to dedicated environmentalists, it suggests the need for a broader multiclass, multiarea movement for ecological repair, a strategy that moves from the pursuit of localized political remedies by compositional elites to the quest for political actions involving the activation and empowerment of local citizens in environmental disputes, a program that bridges class interests and relates ecological repair to democratic economic planning, and an orientation that goes beyond the "not-in-my-backyard"/obstructionist view of the past to a positive, reconstructive posture for the future offering new ways of promoting the economic livelihood and social well-being of our citizens within the bounds of the ecosystem (Yanarella, 1985).

REFERENCES

Anderson, Frederick. 1973. *NEPA and the Courts: A Legal Analysis of the National Environmental Policy Act*. Washington, D.C.: Resources for the Future.

Ashley, David. Interview. July 11, 1982. Georgetown, Kentucky.

Bellah, Robert N.; Madsen, Richard; Sullivan, William M.; Swidler, Ann; and Tipton, Steven M. 1985. *Habits of the Heart: Individualism and Commitment in American Life*. Berkeley: University of California Press.

Berger, Peter, and Richard John Neuhaus. 1977. *To Empower People: The Role of Mediating Structures in Public Policy*. Washington, D.C.: American Enterprise Institute.

Bishop, A. Bruce. 1975. "Public Participation in Environmental Impact Assessment." In *Environmental Impact Assessment*, ed. by Marlan Blissett. Austin, Tex.: Engineering Foundation, pp. 219–36.

Boyte, Harry. 1980. *The Backyard Revolution: Understanding the New Citizen Movement*. Philadelphia, Pa.: Temple University Press.

Burton, Dudley J. 1980. *The Governance of Energy*. New York: Praeger Publishers.

Caldwell, Lynton; Lynton R. Hayes; and Isabel M. MacWhirter. 1976. *Citizens and the Environment*. Bloomington, Ind.: Indiana University Press.

Cappello, Joseph. March 18, 1977. Letter to Tom Lohr, Irvin Vice President.

"Changes in U.S. Energy Plans Slow Coal-Gas Development, Official Says." March 22, 1979. *Louisville Courier-Journal*, p. 14.

Cloward, Richard A., and Frances F. Piven. 1979. *Poor People's Movements: Why They Succeed, How They Fail*. New York: Random House.

Cohn, Ray. "State Official Played Dual Role in Scott Gasifier Project." November 12, 1978. *Lexington Herald-Leader*, pp. B–1, B–4.

Congressional Research Service. 1976. *Federal Energy Organization: Historical Perspectives*. Washington, D.C.: Government Printing Office.

"Damper Is Placed on Kentucky's Hopes for Coal Conversion." January 23, 1979. *Louisville Courier-Journal*, p. 5.

Daneke, Gregory A. "A Strategic Planning Approach to Energy/Environment Tradeoffs." 1980. Tempe, Ariz.: University of Arizona, Public Policy Planning and Administration.

Drake, David. October 25, 1976. Letter to Tom Lohr, Irvin Vice President.

———. November 17, 1976. Letter to Tom Lohr, Irvin Vice President.

———. June 3, 1977. Letter to Tom Lohr, Irvin Vice President.

———. July 27, 1977. Letter to Tom Lohr, Irvin Vice President.

"Gasification Plant Development Discussed at Council Meeting." October 13, 1976. *Georgetown Times*, p. 1.

Gellhorn, Ernest. 1975. "Public Participation in Administrative Proceedings." In *Environmental Impact Assessment*, ed. by Marlan Blissett. Austin, Tex.: Engineering Foundation, pp. 191–217.

Goodwin, Craufurd D., ed. 1981. *Energy Policy in Perspective: Today's Problems, Yesterday's Solutions*. Washington, D.C.: Brookings Institution.

Graddy, W. Henry. July 16, 1982. Interview. Versailles, Kentucky.

Hahn, O. J. June 10, 1982. Interview. Lexington, Kentucky.

Hall, Betty Ann. June 28, 1982. Interview. Georgetown, Kentucky.

Horwitch, Mel. 1979. "Coal: Constrained Abundance." In *Energy Future: Report of the Energy Project of the Harvard Business School*, Ed. by Robert Stobaugh and Daniel Yergin. New York: Random House, pp. 100–134.

Horwitch, Mel, and Vietor, Richard H. K. July 1980. "The Political Management of Synthetic Fuels: A Retrospective Appraisal." Unpublished paper.

Humphrey, Craig R., and Buttel, Frederick R. 1982. *Environment, Energy, and Society*. Belmont, Calif.: Wadsworth Publishing Company.

Joint Committee on Atomic Energy. 1975. *ERDA Authorizing Legislation, FY 1976*. Hearings before the JCAE, pt. 1. Washington, D.C.: Government Printing Office.

Kentucky Center for Energy Research. June 1976. *A Perspective for Energy Development Policy Planning in Kentucky*. Lexington, Ky.: KCER.

———. December 26, 1976. Staff Memorandum on the Draft Environmental Assessment for the Kentucky-Federal Energy Park.

Kentucky Energy Research Board. October 19, 1976. Minutes.

Leahy, Peter J., and Mazur, Allan. August 1980. "The Rise and Fall of Public Opposition in Specific Social Movements." *Social Studies of Science* 10:259–80.

Lohr, Tom. April 9, 1976. Letter to David Drake, KCER Administrator.

———. April 20, 1976. Letter to Jack Kilmeyer, Manager, General Electric—Lamp Division.

———. June 10, 1976. Letter to Jackson White, Secretary of the Cabinet.

———. October 6, 1976. Letter to David Drake, KCER Administrator.

———. December 1, 1976. Letter to Russell Bardos, ERDA Project Manager.

———. October 7, 1977. Letter to David Drake, KCER Administrator.

Mason & Hanger-Silas Mason Co., Inc. January 1976. *Environmental Assessment for the Kentucky-Federal Energy Park, Georgetown, Kentucky*.

Mazur, Allan. Spring 1975. "Opposition to Technological Innovation." *Minerva* 13: 58–81.

———. 1981. *The Dynamics of Technical Controversy*. Washington, D.C.: Communications Press.

Mitchell, John. June 1, 1982. Interview. Lexington, Kentucky.

Neiman, Max, and Loveridge, Ronald O. November 1981. "Environmentalism and Local Growth Control: A Probe into the Class Bias Thesis." *Environment and Behavior* 13:759–72.

"Offutt Announces for Council Re-election." April 18, 1979. *Georgetown News & Times*, p. 10.

Phares, Nancy Nunnelly. June 14, 1982. Interview. Lexington, Kentucky.

Proceedings of a Public Hearing on a Proposed Kentucky-Federal Energy Park in Scott County, Kentucky. Garth Elementary School, Georgetown, Kentucky. January 27, 1977. Transcript. Georgetown: Georgetown Public Library.

Proceedings of a Public Hearing on Low BTU Coal Gasification Facility and Industrial Park, Georgetown, Kentucky [DEIS, DOE/EIS–000–7–D]. John L. Hill

Chapel, Georgetown College, Georgetown, Kentucky. August 22, 1978. Transcript. Georgetown: Georgetown Public Library.

Reid, Herbert G., and Yanarella, Ernest J. 1981. "Beyond the Energy Complex: The Role of the Public Sphere." Paper prepared for delivery at the 1981 annual meeting of the American Political Science Association, September 3–6, 1981, New York City.

Rosenbaum, Walter A. 1973. *The Politics of Environmental Concern*. New York: Praeger Publishers.

———. 1981. *Energy, Politics, and Public Policy*. Washington, D.C.: Congressional Quarterly.

Schattschneider, E. E. 1960. *The Semisovereign People: A Realist's View of Democracy in America*. New York: Holt, Rinehart and Winston.

Snyder, J. Robert. July 12, 1982. Interview. Lexington, Kentucky.

Teich, Albert. 1977. "Bureaucracy and Politics in Big Science: Relations between Headquarters and the National Laboratories in AEC and ERDA." Unpublished paper.

Thrailkill, John. June 17, 1982. Interview. Lexington, Kentucky.

Traynor, Harry. December 6, 1976. Memorandum for KCER Files.

———. January 13, 1977. Memorandum to David Drake, KCER Administrator.

———. October 10, 1977. Memorandum to David Drake, KCER Administrator.

U.S. Congress, Office of Technology Assessment. 1975. *An Analysis of the ERDA Plan and Program*. Washington, D.C.: Government Printing Office.

U.S. Department of Energy. April 1978. *Low BTU Coal Gasification Facility and Industrial Park, Georgetown, Kentucky*. DEIS, DOE/EIS–000-7-D. Washington, D.C.: U.S. DOE.

U.S. Energy Research and Development Administration (ERDA). April 13, 1976. Pre-Selection Conference on PON-FE–4. Washington, D.C.

———. October 12, 1976. Letter to John J. Geel, Irvin President.

Vietor, Richard H. K. Spring 1980. "The Synthetic Liquid Fuels Program: Energy Politics in the Truman Era." *Business History* 54(1):1–34.

Wellstone, Paul David. 1978. *How the Rural Poor Got Power: Narrative of a Grass-Roots Organizer*. Amherst: University of Massachusetts Press.

Wenner, Lettie McSpadden. 1976. *One Environment under Law: A Public Policy Dilemma*. Pacific Palisades, Calif.: Goodyear Publishing Company.

———. 1982. *The Environmental Decade in Court*. Bloomington, Ind.: Indiana University Press.

Wolfe, Alan. 1977. *The Limits of Legitimacy*. New York: Basic Books.

Yanarella, Ernest J. 1985. "Environmental vs. Ecological Perspectives on Acid Rain: The American Environmental Movement and the West German Green Party." In *The Acid Rain Debate: Scientific, Economic, and Political Dimensions*, ed. by Ernest J. Yanarella and Randal H. Ihara. Boulder, Colo.: Westview Press, pp. 243–60.

Yanarella, Ernest J., and Randal H. Ihara. 1978. "The Military/Energy Connection: The Institutionalization of the 'Technological Breakthrough' Approach to Energy R&D." *Northeast Peace Science Review* 1:187–207.

7
Public Ambivalence about Synthetic Fuels and Other New Energy Development

CYNTHIA M. DUNCAN AND ANN R. TICKAMYER

Synthetic fuel development vividly illustrates the dilemma that individuals and communities face in assessing the costs and benefits of new economic activity. While resource development can provide jobs and revenue for host communities, extraction and processing invariably entail environmental and social costs for resource areas. These costs can be borne by the project, leading either to higher costs for consumers or lower profits for capital, or they can be borne by the community and workers in the form of increased expenses for infrastructure and local services, pollution, and potential health hazards. The allocation of these costs is part of a complex political process that rests partly on how people perceive the impact of the costs on the community and partly on whether they believe they have economic alternatives.

In an effort to understand how these decisions are made, social scientists have conducted numerous public-opinion surveys, many in conjunction with socioeconomic impact studies mandated by new energy development projects. Despite a profusion of such studies, scholars have had little success in explaining public attitudes on the topic. In this chapter we explore reasons for this failure by reexamining the results from past work and by presenting our own recent research. In particular, we critically reevaluate a representative selection of studies from western states in conjunction with the findings from our study of public opinion on the costs and benefits of synthetic fuel development in Kentucky. Comparison of cross-regional similarities and dif-

An earlier version of this paper was presented at the annual meeting of the Rural Sociology Society, College Station, Texas, August, 1984. The authors contributed equally. Thomas Ford and Timothy Johnson provided helpful advice on construction of the survey, and Kathleen Blee, William Duncan, and Thomas Rudel were good critics of an early draft.

ferences leads us to a reformulation of current theory and research on attitudes toward development.

BACKGROUND THEORY AND RESEARCH

Revising the Growth Machine Thesis

Public understanding of the costs of large-scale energy development and the allocation of these costs has been changing over the last fifteen years. For over a century, energy development in the United States, like industrial development in general, took place within an atmosphere of community boosterism (Molotch, 1976; McNulty, 1978; Pierce et al., 1979; Banks, 1979; Eller, 1982; Cobb, 1982). Community elites and nonelites, workers and public officials, assumed that growth in local economic activity would benefit the area, providing jobs, income, and new tax revenues. Residents valued these economic benefits so much that they rarely took external costs into account (Anshen, 1970, p. 9). Growth was accepted as an unqualified advantage, and its effect on the quality of life and the environment was not questioned. But over the last two decades it has become clear that the "growth machine" mentality (Molotch, 1976) no longer accounts completely for attitudes toward development (Summers et al., 1979; Ballard et al., 1981; Barrows and Charlier, 1982; Little and Krannich, 1982).

Confidence in the benefits of unfettered economic growth began to waver in the 1960s. Over the next few years business and policy analysts pointed to challenges to the implicit "social contract" between the private and public sector. They cited racial unrest, neighborhood organizing, Vietnam protests, and the growth of the environmental movement as evidence that Americans had begun to distrust business and government. The public appeared to be demanding more social responsibility from corporations (Anshen, 1970; Post, 1970; Rawls, 1971; Jacoby, 1973; McKie, 1974; Vogel, 1978; Beauchamp and Bowie, 1983; Douglas and Wildavsky, 1982; Friedland, 1983). The National Environmental Policy Act of 1970 is an example of legislation that institutionalized these new attitudes about responsibility for costs of production. The Act gave citizens the right to force private developers to take account of external costs and included provisions for citizen participation in decisions about meeting those costs (McEvoy, 1972; Buttel and Flinn, 1974; Utton et al., 1976; Duerksen, 1983).

When the Arab oil embargo of the early 1970s prompted new, large-scale energy development by multinational corporations in western states, close attention was paid to the costs as well as the benefits for local resource areas (Krutilla et al., 1978; Leistritz and Murdock, 1981; Weber and Howell, 1982). Since these projects were large and often occurred in remote areas with limited public services, existing local communities became boomtowns that faced enormous public costs (Meeker, 1976; U.S. Senate, 1979; Stinson et al., 1982). Moreover, some of these new development projects occurred in states that had experienced heavy costs from other extraction industries (Watkins, 1982; Power, 1983). Environmental and community activists in western states were able to garner the public support and political strength necessary to require energy developers to bear more of the costs of development (Duncan, 1985). In a 1982 report titled *Mitigating Socioeconomic Impacts of Energy Development* the U.S. General Accounting Office (GAO) reported substantial industry assistance to cover costs associated with new developments in Rocky Mountain states (GAO, 1982, pp. 33–35). In summary, support for unrestricted growth and development has seen increasing opposition from a variety of sources.

Review of Western Opinion Studies

To help prepare for the substantial economic and social changes that would accompany new projects in western states during the late 1970s, governments responsible for public well-being in isolated rural areas prepared socioeconomic impact studies. Public opinion surveys were frequently included in these studies to assess both local citizens' favorability toward development projects, based on their perceptions of the benefits, and their concerns about potential disadvantages, based on their assessment of the costs. We review seven representative surveys conducted in western states in the late 1970s, including Selbyg (1978), Lopreato and Blissett (1977), Faulkner and Howard (1975), Mid-Continent Surveys (1978), Thompson and Blevins (1983), Lewis and Albrecht (1977), and Ludtke (1977). In addition, the composite tables constructed by Murdock and Leistritz (1979) to compare survey results were examined. A literature review identified opinion surveys that tapped attitudes toward new energy development, and these seven studies were chosen for closer examination because they probed for differences in opinion and, as a group, represented geographical diversity. Surveys were conducted in parts of North Dakota, Montana, Wyoming, Nebraska, Idaho, Utah, and Texas.

Several common issues emerge from this review. First, researchers repeatedly find that the majority of respondents favor development projects because they expect economic benefits. However, the margins of favorability are reduced when questions of social and environmental costs arise. While the focal points and presentations vary considerably in these studies, they consistently conclude that overall respondents expect economic benefits but are cautious about potential costs.

Second, there is surprisingly little variation across subgroups in the populations studied. Most studies report negative findings from subgroup differences, including sociodemographic differences such as age, sex, income, and education, as well as occupational differences. Few differences in opinion occurred even between groups that had already experienced development and those that had not (Selbyg, 1978; Lopreato and Blissett, 1977; Faulkner and

Howard, 1975; Mid-Continent Surveys, 1978; Thompson and Blevins, 1983; Lewis and Albrecht, 1977; Ludtke, 1977).

Third, closer examination of studies that do describe differences across subgroups or sociodemographic characteristics raises questions about the importance of those findings. Many studies of western energy development do not provide detailed empirical analysis (Wilkinson et al., 1982; Murdock and Leistritz, 1979, p. 255). Most were originally commissioned by government bodies and are geared toward lay audiences; consequently, they do not report statistical findings (Faulkner and Howard, 1975; Ludtke, 1977). Others are impressionistic accounts or ethnographic case studies and, therefore, have no statistical findings to report (Cortese and Jones, 1977; Little, 1977; Gold, 1974).

Finally, studies that describe results of statistical analyses frequently use bivariate measures and only report results of significance tests. This results in an emphasis on statistically significant findings with little corresponding attention to their size or substantive significance. Our calculations suggest that their explanatory power is usually very small, even compared to other attitude surveys (Swanson and Maurer, 1983, p. 654). For example, Thompson and Blevins (1983) report the results of a series of one-way analyses of variance F tests examining community differences in indices of respondents' attitudes about economic opportunity, social change, and environmental concern. All exhibit statistical significance reflecting differences between boomtown and unaffected communities. Closer inspection of these results shows that the differences are very small. Using their reported findings to construct Eta square measures of association (equivalent to R square for ANOVA tables), we find the Eta square for economic opportunity is .023, .046 for social change, and .008 for environmental concerns. Similarly, Lewis and Albrecht's findings report statistically significant differences in attitudes of residents from a more economically prosperous county and a poorer one, but the overall explanatory power is still less than 6 percent in every case.

A recent national comparative study of favorability toward nuclear power plants by Freudenburg and Baxter (1983) is an exception. Their study explains a remarkable 70 percent of the variation in host community favorability toward local nuclear facilities when they use a dummy variable representing "before" and "after" the Three Mile Island accident. Importantly, what they explain appears to be reactions to the potential costs of such development. Since the accident at Three Mile Island occurred, more people appear to see the costs of nuclear power development as potentially life-threatening. They know they do not want to accept the risks associated with hosting a nuclear power plant.

While most researchers separate questions of costs, the "social impact," from benefits, the "economic impact," some do not recognize that the same individuals can see development as simultaneously beneficial and costly. For example, Freudenburg (1983) distinguishes between those who view development as bringing economic benefits and those who regard its negative impact on the quality of life, but he treats these as exclusive alternative views in order to test competing hypotheses. We argue here that most people see both sides and are confused when they weigh them both.

When costs are less clearly defined, residents are presented with a dilemma in which the economic benefits are comprehensible, but the costs and potential negative consequences to their quality of life are essentially unknown. In one way or another, Ludtke (1977), Selbyg (1978), Lopreato and Blissett (1977), and Murdock and Leistritz (1979) conclude that, overall, people find the issues complex, and in many cases their uncertainty leads them to an undecided position. Selbyg points out that this indecision appears in survey results as inconsistent and unexplainable responses (1978, pp. 112, 132).

RESEARCH DESIGN

When large-scale oil-shale projects were proposed in areas adjacent to the coal fields in the eastern United States, they provided an opportunity to compare public opinion regarding new energy development in the East with that in the West. To investigate attitudes toward energy development in an eastern state, we developed a questionnaire that examined how residents regard the costs and benefits such projects entail. Two telephone polls of residents in three eastern Kentucky counties were conducted in July and August of 1983. One sample of 402 respondents was interviewed in Montgomery County with a response rate of 74.4 percent, and another sample of 403 respondents was interviewed in Rowan and Fleming counties with a response rate of 67.9 percent. Subsequently, Rowan County was dropped from the analysis because further investigation indicated that the project would have an impact primarily upon Fleming County. Preliminary analysis showed few differences between Rowan County residents and other respondents. Fleming County respondents were retained for analysis and comparison with Montgomery County residents. We also dropped cases where respondents refused to answer or said they "did not know" over half the time.

On the one hand, communities in eastern coal states always have borne heavy social and environmental costs (Gray, 1933; Caudill, 1962; Bowman and Haynes, 1963; Smith, 1981; Simon, 1981). Workers have experienced long and frequent layoffs and communities have put up with polluted streams and water systems, deteriorating roads, coal dust, and crowded, substandard houses (Balliett, 1978; Duncan and Duncan, 1983; Seltzer, 1985). Because of this history of apparent acceptance of the costs of development (Gaventa, 1980), one might expect less ambivalence about costs of new energy development among eastern residents.

On the other hand, these new shale developments were being proposed at a time of growing national concern for the environment and increasing awareness of community costs associated with development. Although the proposed sites of the shale development were areas that would welcome job development, they were not already dependent upon an extractive economy. These factors suggested that residents might share the cautious attitudes exhibited in some western areas, and thus public opinion in eastern areas might mirror that in western states.

Independent variables included the effect of a number of social and demographic variables that have been important in past research or were of theoretical interest. We examined educational attainment (EDUC), age (AGE), sex (SEX), family income level (INCOME), occupation of household head (OCCUP), and whether or not the family owned property in the county (OWNPROP). We also examined the effect of knowledge about the project (HEARD).[1] In addition to characteristics of respondents, county of residence (COUNTY), was chosen as an independent variable. Fleming and Montgomery counties are predominantly rural and have similar socioeconomic conditions. Poverty levels and unemployment rates, labor participation rates, and per capita income in both counties are below both state and national standards. The counties differed, however, on the amount of local information available on the oil-shale projects and on levels of local activism.[2]

Dependent variables were drawn from the literature on energy development, environmental issues, and quality of life. After examining individual items, we constructed a general attitude toward energy development scale consisting of all the dependent variables (GENATT) and factor scales that represent four different dimensions of public opinion on these issues: the respondents' expectations of economic opportunity (ECONOPP), opinions about the projects' cost to the quality of life in the community (QOLCOST), attitudes toward government involvement in the projects (GOVTATT), and confidence in the developers (COMPATT).[3] (Individual items and scale statistics are found in the Appendix.) These scales were dependent variables in multivariate analyses that assessed the impact of information and activism as well as individual sociodemographic traits that past research has shown to influence attitudes toward energy development. These results were then used to make comparisons between our findings and similar western public opinion surveys and to suggest a reconceptualization of development issues for future research.

RESULTS

Initial bivariate analysis indicates a general tendency for respondents to be optimistic across all categories of the independent variables. The vast majority expect oil-shale development to benefit their county, although there is less optimism on items related to costs and the impacts of development on the quality of life (Table 7–1). While we identify several factors that influence attitudes on economic opportunities that accompany oil-shale development, background characteristics taken individually do not explain the variation in

Table 7–1
Contingency Tables of Attitudes toward Oil-Shale Development by Individual Background Characteristics

	INFORMED		EDUCATION			FAMILY INCOME			SEX	
	Heard N=535	Not Heard N=138	Under High School N=254	High School N=263	Over High school N=167	Under $10,000 N=152	$10,000-20,000 N=213	Over $20,000 N=232	Male N=325	Female N=362
Economic Impacts:										
BENEFCTY	63.4	71.8	66.2	67.1	61.8	69.2	72.1	61.9	65.1	65.3
YNGSTAY	61.9**	74.0**	72.9***	62.3***	54.8**	75.5***	65.5***	57.1***	62.0	66.2
GOODBUS	86.7**	94.9**	90.1	89.6	86.5	90.1	89.6	86.5	87.4	88.9
BRINGTAX	89.6*	96.4*	92.3	91.0	91.1	92.3	91.0	91.1	91.7	90.0
TAXIMPROV	66.1**	80.5**	74.4	65.3	69.2	80.9***	72.5***	59.2***	65.6*	72.8*
Quality of Life Impacts:										
SAMECRIM	56.3	64.4	52.3	61.3	61.0	55.6	64.2	57.1	58.8	57.3
LOCALDEC	57.3	56.1	57.2	53.5	61.3	52.0	61.5	57.1	56.7	56.9
FRIENDLY	61.8	61.8	58.8	65.2	63.5	63.9	65.5	59.6	56.2**	68.5**
PAYROADS	37.2	42.0	40.2	42.1	31.0	39.8	43.2	34.9	39.6	37.2
ADEQWATR	40.3	41.8	41.2	40.4	38.8	35.7*	49.7*	38.8*	44.4*	36.1*
NOTAXINC	38.5**	23.3**	30.9**	34.7**	47.0**	28.0**	36.1**	45.2**	43.5***	28.7***
Approve Government Help:										
GOVTHELP	46.2*	58.6*	52.0*	51.5	40.3*	51.7*	56.5*	42.9*	47.6	49.7
SFCHELP	48.5	55.3	48.7	51.0	50.7	56.1	54.9	47.6	51.9	48.2
STATHELP	58.3	60.7	57.3	62.0	57.7	61.1	65.0	58.6	60.1	57.5
Confidence in Company:										
TRUSTCO	49.3	57.5	47.6	53.0	51.4	50.0	55.1	49.5	51.1	50.4
TRUSTFIX	51.7	49.2	45.8	54.7	54.7	50.4	55.3	55.6	52.6	50.3
GOODLEAS	55.8*	72.7*	59.4	56.6	60.8	60.7	64.4	54.6	56.1	61.1

Note: See the appendix for explanation of abbreviation for dependent variables

*$p \leq .05$
**$p \leq .01$
***$p \leq .001$

Table 7–1 (Continued)

	AGE			OWN PROPERTY		COUNTY	
	Under 31 yrs N=166	31-50 yrs. N=268	Over 50 yrs. N=268	No, Don't N=142	Yes, Do N=529	Montgomery N=402	Fleming N=285
Economic Impacts:							
BENEFCTY	75.9**	64.4**	58.3**	62.7*	75.2*	65.2	65.2
YNGSTAY	71.4	60.2	63.6	61.2**	75.0**	59.6***	70.2**
GOODBUS	92.9	87.1	86.1	87.5	91.1	87.4	89.2
BRINGTAX	92.3	91.2	89.4	90.3	92.7	88.5**	94.3**
TAXIMPROV	76.1**	62.6**	72.2**	66.8**	78.3**	64.1***	76.5***
Quality of Life Impacts:							60.6
SAMECRIM	66.9***	64.3***	43.8***	57.7	57.8	56.3	51.8*
LOCALDEC	55.1	62.4	51.8	57.6	53.3	60.3	63.8
FRIENDLY	62.1	66.4	57.8	65.6	62.7	61.4	41.8
PAYROADS	43.7	37.7	35.3	38.2	40.4	36.2	35.8
ADEQWATR	49.3***	44.7***	28.9***	37.7*	50.5*	43.7	34.5
NOTAXINC	31.3	39.5	35.3	37.5	33.6	37.1	
Approve Government Help:						47.1	50.8
GOVTHELP	60.5**	46.2**	42.7**	47.8	52.2	50.0	50.0
SFCHELP	57.7**	52.9**	40.6	48.8	57.1	58.5	59.8
STATHELP	64.6***	64.0***	48.8***	59.3	60.0		
Confidence in Company:						50.7	50.6
TRUSTCO	60.5***	52.1***	41.2***	49.6	54.2	52.0	50.9
TRUSTFIX	54.3*	56.2*	43.6*	50.8	54.6	58.6	58.0
GOODLEAS	66.0	58.2	51.9	56.7	65.7		

*p≤ .05
**p≤ .01
***p≤ .001

opinion on the costs of development, attitudes about government assistance to development projects, or levels of confidence in company responsibility.

Much of the past research on attitudes toward energy development has been restricted to bivariate or very simple multivariate analysis involving three or four variables. To see whether we could increase our understanding of the relationships, we used multivariate techniques. After exploratory analysis eliminated the possibility of significant interactions and nonlinearities, we used ordinary least squares regression to examine the effect of all independent variables on scaled attitudes.

Table 7–2 presents the results of the multivariate analysis for the general scale of attitudes toward energy development (GENATT) as well as for the smaller subset of variables composing an economic impact scale (ECONOPP), a quality of life impact scale (QOLCOST), an attitude toward government aid scale (GOVTATT), and a confidence in oil-shale development scale (COMPATT). Standardized and unstandardized parameters and standard errors of estimates as well as summary measures of explained variance and F ratios are included in this table.

Our results are similar to the research discussed above. Few relationships appear when all predictor variables are examined simultaneously. The largest finding is for the ECONOPP scale. Six percent of the variance in attitudes toward economic impact of oil-shale development is explained by the combined independent variables, with county of residence (COUNTY), education (EDUC), and property ownership (OWNPROP) all having a significant negative impact. In other words, residents of the county with more information and activism, more property owners, and more highly educated respondents are less optimistic about the economic benefits of energy development.

As researchers found in western studies, we explain even less variance in scales that measure more than expected benefits. Three percent of the variance in the general attitudes toward development scale (GENATT) is accounted for, and the only parameter estimate that has a statistically or substantively significant result is AGE. One percent of the variance in attitudes toward the costs of oil-shale development (QOLCOST) is explained, and only AGE attains statistical significance. Three percent of the attitude toward government aid (GOVTATT) is accounted for, and again only AGE has a small positive significant effect. Finally, none of these predictors has explanatory power for people's confidence in the development company (COMPATT).

The results of the multivariate analysis show that county and individual factors contribute somewhat to an explanation of attitudes toward economic benefits of oil-shale development. Most other aspects of attitudes toward energy development cannot be explained by independent variables in this study. The only consistent predictor across all dependent variables is AGE. Older people are less positive about all facets of energy development.

Table 7–2
Regression of Attitude toward Energy-Development Scales on Selected Independent Variables (N = 601)

Independent Variables	GENATT		ECONOPP		QOLCOST		GOVTATT		COMPATT	
	b	Beta	b	Beta	b	Beta	b	Beta	b	Beta
COUNTY	254 (.351)	.030	-.283* (.120)	-.097	-.122 (.141)	-.036	.052 (.105)	.021	.041 (.092)	.019
HEARD	-.642 (.457)	-.061	-.289 (.156)	-.080	.017 (.184)	.004	-.194 (.137)	-.062	-.176 (.119)	-.065
EDUC	-.039 (.062)	-.030	-.050* (.021)	-.111	.019 (.025)	.036	-.024 (.018)	-.062	.016 (.016)	.047
AGE	.031* (.011)	.124	.007 (.004)	.077	.011* (005)	.107	.011* (.003)	.143	.003 (.003)	.045
OCCUP	.398 (.364)	.047	.063 (.124)	.021	.104 (.146)	.031	.157 (.109)	.062	.074 (.095)	.034
OWN PROP	-.517 (.422)	-.051	-.357* (.151)	-.101	.025 (.178)	.006	-.030 (.132)	-.010	-.156 (.116)	-.059
INCOME	-.000 (.000)	-.028	-.000 (.000)	-.034	-.000 (.000)	-.037	-.000 (.000)	-.029	-.000 (.000)	-.028
SEX	-.287 (.339)	-.035	-.196 (.116)	-.069	.073 (.136)	.022	-.072 (.101)	-.029	-.092 (.089)	-.043
CONSTANT	9.946		4.677		2.116		1.471		1.683	
F^2	2.661*		5.005*		1.249		2.911		1.458	
R^2	.035		.064		.017		.038		.019	
$\bar{R}^2$	.022		.051		.017		.025		.006	

*P ≤ .05

DISCUSSION AND CONCLUSION

The overwhelming evidence from studies of synthetic-fuel and other new energy development is that residents are ambivalent when they weigh the costs of development. Our results suggest that respondents in our eastern oil-shale regions, like westerners, are neither wholeheartedly pro-growth nor anti-growth. They regard the costs and the benefits associated with energy development as distinct issues. They do not want to forgo economic benefits, but they want to avoid high community costs. We draw this conclusion for two reasons. In every case, our analysis shows that we can explain more about attitudes toward the benefits of energy development than we can about attitudes toward the costs. Attitudes toward benefits may reflect underlying political dispositions that are attributable to sociodemographic differences. Attitudes toward costs appear to resist these conventional explanations. Furthermore, benefit and cost items group separately in our factor analysis. The benefit factors group into a scale that has good statistical reliability. However, the items that refer to quality-of-life costs do not hold together as well, and the scale based on them has much lower reliability. If people were committed either to promoting growth or halting it, there would have been more correspondence between the results of the analysis of attitudes toward costs and the analysis of attitudes toward benefits.

Close observers of the debate over oil-shale development in one of our study areas speculate that many residents resist having to choose between the benefits of economic growth and the costs to the environment and community (Lamm and Murphy, 1983).[4] They want the jobs and so are pro-development; they are worried about the impacts and so appear to be antidevelopment. Development issues are complex for residents, and their responses to questions about benefits and costs reflect that complexity and their own uncertainty.

In western states this uncertainty has been resolved partially through negotiations with developers over the distribution of costs. Socioeconomic impact studies and mitigation agreements are common practice in western energy communities (Power, 1983; Weber and Howell, 1982; Murdock and Leistritz, 1979; Sullam et al., 1978). Given the similarity in attitudes toward costs and benefits between residents in western and eastern states, the question arises whether people in the eastern energy-producing region would support a more aggressive community stance toward developers. Future analysis should investigate public support for negotiations over impact mitigation in eastern energy development areas, and should make direct comparisons with western communities.

This analysis of public opinion of synthetic-fuel development found widespread uncertainty toward costs and benefits. The authors believe that this public ambivalence extends beyond synthetic-fuel development. Researchers who analyze public attitudes toward energy development in particular and industrial development in general need to reconceptualize their research questions. They need to take account of the uncertainty about potentially dangerous costs that prevails across social class and economic circumstances. Researchers need to consider how the general public and particular social classes calculate the trade-offs presented by development.

NOTES

1. People who had not heard of the project were more likely to say that public information about the project was inadequate. Although 80 percent had heard of oil-shale development in their county, only 6 percent knew the company's name or location.

2. The Montgomery County paper carried 57 stories about the Means Project between January 1981 and November 1983, while the two papers in Fleming County combined carried only 12 stories. The Montgomery paper had stories discussing potential fiscal or environmental problems and stories about western boomtowns and occasionally devoted its editorial page to letters on the issue. The Fleming papers rarely mentioned opposition to shale development and had few letters from readers on the subject, either pro or con (Brewster, 1983). Montgomery County had both an "official" citizens commission on oil-shale development appointed by the county judge and an unofficial grassroots citizens group. Although a citizens group had been formed in Fleming County when the oil-shale issue first emerged, it was abandoned.

3. Most analysts have not assessed attitudes toward these relationships directly (see Murdock and Leistritz, 1979; Stinson et al., 1982; Sullam et al., 1978). However, Lopreato and Blissett asked Texas residents whether the government or private companies should develop geothermal test wells. A majority (59 percent) was in favor of private-sector development, and a majority (62 percent) agreed that their community should receive funding and advice from governments (1977, p. 28).

4. Without this survey, observers would have concluded that the county was sharply polarized. Lamm points out that undecided citizens have no avenue for expression in the debate (Lamm, 1984). Our results indicate that most researchers were undecided and uncertain about the project.

REFERENCES

Anshen, Melvin. November–December 1970. "Changing the Social Contract." *Columbia Journal of World Business* 5:6–14.

Baldwin, Thomas E.; Dixon-Davis, Diane; Stenehjem, Erik J.; and Walsko, Thomas D. 1976. *A Socioeconomic Assessment of Energy Development in a Small Rural County: Coal Gasification in Mercer County, North Dakota*. Argonne, Ill.: Argonne National Laboratory.

Ballard, Chester C.; Hoskins, Myrna S.; and Copp, James H. 1981. *Local Leadership Control of Small Community Growth*. Technical Report no. 81–3. College Station: Texas Agricultural Experiment Station, The Texas A & M University System.

Balliett, Lee. 1978. "A Pleasing Tho Dreadful Sight: Social and Economic Impacts of Coal Production in the Eastern Coalfields." Report to the Office of Technology Assessment, U.S. Congress.

Banks, Alan. 1979. "Labor and the Development of Industrial Capitalism in Eastern Kentucky, 1870–1930." Ph.D diss., McMaster University, Ontario, Canada.

Barrows, Richard, and Charlier, Marj. 1982. "Local Government Options for Managing Rapid Growth," in *Coping with Rapid Growth in Rural Communities*, ed. by Bruce A. Weber and Robert E. Howell. Boulder, Colo.: Westview Press, pp. 193–220.

Beauchamp, Tom L., and Bowie, Norman, eds. 1983. *Ethical Theory and Business*. Englewood Cliffs, N.J.: Prentice-Hall.

Bowman, Mary Jean, and Haynes, W. Warren. 1963. *Resources and People of East Kentucky: Problems and Potentials of a Lagging Region*. Baltimore: Johns Hopkins University Press.

Brewster, Rosa. 1983. Memorandum to Mountain Association for Community Economic Development, reporting news coverage of oil shale in Fleming and Montgomery counties. Berea, Ky.

Buttel, Frederick, and Flinn, William. Spring 1974. "The Structure of Support for the Environmental Movement, 1968–1970." *Rural Sociology* 39:55–69.

Caudill, Harry M. 1962. *Night Comes to the Cumberlands: A Biography of a Depressed Area*. Boston: Little, Brown and Company.

Clemente, Frank A., and Krannich, Richard S. 1982. "Energy," in *Rural Society in the U.S.: Issues for the 1980's*, ed. by Don A. Dillman and Daryl J. Hobbs. Boulder, Colo.: Westview Press, pp. 34–43.

Cobb, James C. 1982. *The Selling of the South: The Southern Crusade for Industrial Development, 1936–1980*. Baton Rouge: Louisiana State University Press.

Cortese, Charles F., and Jones, B. 1977 "The Sociological Analysis of Boom Towns." *Western Sociological Review* 8:76–90.

Douglas, Mary, and Wildavsky, Aaron. 1982. *Risk and Culture: An Essay on the Selection of Technical and Environmental Dangers*. Berkeley: University of California Press.

Duerksen, Christopher J. 1983. *Environmental Regulations of Industrial Plant Siting: How to Make It Work Better*. Washington, D.C.: Conservation Foundation.

Duncan, Cynthia L. 1985. "Capital and the State in Regional Economic Development: The Case of the Coal Industry in Central Appalachia." Ph.D. diss., University of Kentucky.

Duncan, Cynthia L., and Duncan, William. 1983. "Coal, Poverty, and Development Policy in Eastern Kentucky," in Rural Development, Poverty, and Natural Resources Workshop Paper Series, Part 5. Washington, D.C.: National Center for Food and Agricultural Policy, Resources for the Future, pp. 25–46.

Eller, Ronald D. 1982. *Miners, Millhands, and Mountaineers: Industrialization of the Appalachian South, 1880–1930*. Knoxville: University of Tennessee Press.

Faulkner, Lee, and Howard, Mike. 1975. *Private Opinions and Public Decisions*. Bozeman: Montana University Joint Water Resources Research Center.

Friedland, Roger. 1983. "The Politics of Profit and the Geography of Growth." *Urban Affairs Quarterly* 19(1):41–45.

Freudenburg, William R. 1983. "Boomtown's Youth: The Differential Impacts of Rapid Community Growth on Adolescents and Adults." *American Sociological Review* 49:697–705.

Freudenburg, William R., and Baxter, Rodney K. 1983. "Public Attitudes toward Local Nuclear Power Plants: A Reassessment." Paper presented at the 1983 annual meetings of the American Sociological Association, Detroit, Michigan.

Gaventa, John. 1980. *Power and Powerlessness: Quiescence and Rebellion in an Appalachian Valley*. Urbana: University of Illinois Press.

Gilmore, J. S., and Duff, M. K. 1975. *Boom Town Growth Management: A Case Study of Rock Springs–Green River, Wyoming*. Boulder, Colo.: Westview Press.

Gold, Raymond L. 1974. *A Comparative Case Study of the Impact of Coal Development in the Way of Life of People in the Coal Areas of Eastern Montana and Northeastern Wyoming*. Denver: Northern Great Plains Resources Program.

Gray, L. C. July 1933. "Economic Conditions and Tendencies in the Southern Appalachians as Indicated by the Cooperative Survey." *Mountain Life and Work*, pp. 7–12.

Howell, Robert E., and Weber, Bruce A. 1982. "Impact Assessment and Rapid Growth Management." In *Coping with Rapid Growth in Rural Communities*, ed. by Bruce A. Weber and Robert E. Howell. Boulder, Colo.: Westview Press, pp. 243–68.

Jacoby, Neil. 1973. *Corporate Power and Social Responsibility*. New York: Macmillan.

Kneese, Allen. 1985. *Measuring the Benefits of Clean Air and Water*. Baltimore: Johns Hopkins University Press.

Krutilla, John V., and Fisher, Anthony C., with Rice, Richard E. 1978. *Economic and Fiscal Impacts of Coal Development: Northern Great Plains*. Baltimore: Johns Hopkins University Press.

Lamm, Carol. February 1984. Private communication with authors.

Lamm, Carol, and Murphy, Tim. December 1983. Private communication with authors.

Leistritz, F. Larry and Murdock, Steve H. 1981. *The Socioeconomic Impact of Resource Development: Methods for Assessment*. Boulder, Colo.: Westview Press.

Lewis, C., and Albrecht, S. 1977. "Attitudes toward Accelerated Urban Development in Low-Population Areas." *Growth and Change* 8:22–28.

Little, Ronald L. 1977. "Some Social Consequences of Boomtowns." *North Dakota Law Review* 33(3):401–25.

Little, Ronald L., and Krannich, Richard S. 1982. "Organizing for Local Control in Rapid Growth Communities." In *Coping With Rapid Growth in Rural Communities*, ed. by Bruce A. Weber and Robert E. Howell. Boulder: Westview Press.

Lopreato, Sally Cook, and Blissett, Marlan. 1977. *An Attitudinal Survey of Citizens in a Potential Gulf Coast Geopressured-Geothermal Test-Well Locality*. Washington, D.C.: Energy Research and Development Administration.

Lovejoy, Stephen B. 1977. *Local Perceptions of Energy Development: The Case of the Kaiparowitz Plateau*. Lake Powell Research Project Bulletin, no. 62. Logan: Utah State University.

Ludtke, Richard L. 1977. *Human Impacts of Energy Development: A Survey Study of Dunn, Mercer, and Oliver Counties in North Dakota*. Grand Forks, N.Dak.: Social Science Research Institute.

McEvoy, James, III. 1972. "The American Concern with Environment." In *Social Behavior, Natural Resources and the Environment*, ed. by William R. Burch, Jr., Neil H. Check, Jr., and Lee Taylor. New York: Harper and Row.

McKie, James W. 1974. *Social Responsibility and the Business Predicament*. Washington, D.C.: Brookings Institution.
McNulty, Paul J. Winter 1978. "The Public Side of Private Enterpise: An Historical Perspective on American Business and Government." *Columbia Journal of World Business* 13(4):122–130.
Meeker, D. 1976. *Rapid Growth from Energy Projects: Ideas for State and Local Action*. Washington D.C.: Department of Housing and Urban Development, Office of Community Planning and Development.
Mid-Continent Surveys, Inc. 1978. *Statewide and Regional Attitudes of North Dakota Adults*. Prepared for North Dakota State Planning Division. Bismarck, N.Dak.
Molotch, Harvey. 1976. "The City as a Growth Machine: Toward a Political Economy of Place." *American Journal of Sociology* 82:309–32.
Murdock, Steve H., and Leistritz, F. Larry. 1979. *Energy Development in the Western United States: Impact on Rural Areas*. New York: Praeger Publishers.
Murphy, Timothy. 1983. Private communications with the authors.
O'Hare, Michael; Bacon, Lawrence; and Sanderson, Debra. 1983. *Facility Siting and Public Opposition*. New York: Van Nostrand Reinhold Company.
Petzinger, Thomas, Jr., and Getschow, George. 1984. "Oil's Legacy: Three Part Series on Louisiana and the Oil and Gas Industry." *Wall Street Journal*, October 22, 23, 25.
Pierce, Neal R.; Hagstrom, Jerry; and Steinbach, Carol. 1979. *Economic Development: The Challenge of the 1980's*. Washington, D.C.: The Council of State Planning Agencies.
Post, James E. 1970. *Corporate Behavior and Social Change*. Reston, Va.: Reston Publishing.
Power, Thomas M. November 1983. "Coal Taxation in Montana: A Historical Review and Policy Interpretation." Paper presented at the Kentucky Tax Policy Conference, University of Kentucky, Lexington, Kentucky.
Rawls, John. 1971. *A Theory of Justice*. Cambridge, Mass.: Belknap Press of Harvard University Press.
Selbyg, Arne. 1978. *Residents' Perceptions and Attitudes*. Bismarck, N.Dak.: Regional Environmental Assessment Program.
Seltzer, Curtis. 1985. *Fire in the Hole: Miners and Managers in the American Coal Industry*. Lexington, Ky.: University Press of Kentucky.
Simon, Richard M. Spring 1986. "Uneven Development and the Case of West Virginia: Going beyond the Colonialism Model." *Appalachian Journal* 8:165–86.
Smith, Barbara E. 1981. "Digging Our Own Graves: Coal Miners and the Struggle over Black Lung Disease." Ph.D. diss., Brandeis University, Department of Sociology, Ann Arbor, Mich.: University Microfilms.
Sternleib, George, and Listokin, David. 1981. *New Tools for Economic Development: The Enterprise Zone, Development Bank, and RFC*. New Brunswick, N.J.: Center for Urban Policy Research.
Stinson, Thomas F.; Bender, Lloyd; and Voelker, Stanley W. 1982. *Northern Great Plains Coal Mining: Regional Impacts*. Agriculture Information Bulletin, no. 452. Washington, D.C.: U.S.D.A.
Sullam, Carola; Storper, Michael; Pittman, Donna; and Markusen, Ann. 1978. *Montana: A Territorial Planning Strategy*. Berkeley: Institute of Urban and Regional Development, University of California.

Summers, Gene F.; Beck, E. M.; and Snipp, C. Matthew. 1979. "Coping with Industrialization." In *Nonmetropolitan Industrialization*, ed. by Richard E. Lonsdale and H. L. Seyler. Washington, D.C.: V. H. Winston and Sons, pp. 161–78.

Swanson, Louis E., and Maurer, Richard C. 1983. "Farmers' Attitudes toward the Energy Situation." *Rural Sociology* 48(4):647–60.

Thompson, James G., and Blevins, Audie L. 1983. "Attitudes toward Energy Development in the Northern Great Plains." *Rural Sociology* 48 (1):148–58.

Thompson, J. G.; Blevins, A. L.; and Watts, G. L. 1978. *Socioeconomic Longitudinal Monitoring Report*. Washington, D.C.: Old West Regional Commission.

U.S. General Accounting Office (GAO). 1982. *Mitigating Socioeconomic Impacts of Energy Development*. Gaithersburg, Md.: General Accounting Office.

U.S. Office of Technology Assessment. 1979. *The Direct Use of Coal*. Washington, D.C.: Congress of the United States.

U.S. Senate, Committee on Energy and Natural Resources. 1979. *Energy Impact Assistance: A Background Report*. Hearings, 96th Cong., 1st sess. Washington, D.C.: Government Printing Office.

Utton, Albert E.; Sewell, W. R. Derrick; and O'Riordan, Timothy. 1976. *Natural Resources for a Democratic Society: Public Participation in Decision-Making*. Boulder, Colo.: Westview Press.

Vogel, David. 1978. *Lobbying the Corporation*. New York: Basic Books.

Watkins, Al. August 1982. "North Dakota Populist Seeks Economic Democracy." Interview with Byron Dorgan in *The Texas Observer*, pp. 7–11.

Weber, Bruce A., and Howell, Robert E., eds. 1982. *Coping with Rapid Growth in Rural Communities*. Boulder, Colo.: Westview Press.

Wilkinson, Kenneth P.; Thompson, James G.; Reynolds, Robert R., Jr.; and Otresh, Lawrence M. 1982. "Local Social Disruption and Western Energy Development: A Critical Review." *Pacific Sociological Review* 25(3):275–96.

APPENDIX

Five composite scales were created using factor analytic techniques. Using a principal factors method, a single factor explained almost 80 percent of the variance. The seventeen items with high positive loadings were summed to form our GENATT scale, representing a general set of attitudes toward oil-shale development. (Cronbach's alpha for this scale is .83.) After a varimax rotation, the pattern of loadings suggested subscales corresponding closely to scales found in the literature (Thompson and Blevins, 1983) that made substantive and theoretical sense. Four additional scales were formed by summing their respective items across all cases. ECONOPP represents economic benefits (Cronbach's alpha = .75); QOLCOST reflects attitudes toward the impact on quality of life (alpha = .59); GOVTATT represents attitudes toward government help in oil-shale development (alpha = .77); and COMPATT indicates confidence in the development company (alpha = .55).

Dependent Variables Used in Scales

1. BENEFCTY	whether oil-shale mining will benefit the county overall
2. YNGSTAY	whether increased jobs from oil-shale mining means young people will stay
3. GOODBUS	whether oil-shale mining will be good for businesses
4. BRINGTAX	whether the mining project will bring more tax money
5. TAXIMPRV	whether increased tax money will improve existing community services
6. SAMECRIM	whether the crime rate will stay about the same despite growth
7. LOCALDEC	whether local citizens will maintain control of community decisions
8. FRIENDLY	whether the county will stay friendly with the influx of newcomers
9. PAYROADS	whether the county can afford the cost of increased road use
10. ADEQWATR	whether there is enough water to supply the oil-shale project and county residents
11. NOTAXINC	whether the taxes will stay the same for current residents
12. GOVTHELP	whether an oil-shale mining company should receive financial help from the government
13. SFCHELP	whether this particular company should receive price and loan guarantees from the federal government
14. STATHELP	whether this particular company should receive a grant from the state government to run engineering or environmental tests
15. TRUSTCO	whether the company can be trusted to fulfill its obligations
16. TRUSTFIX	whether the company can be trusted to properly restore the strip-mined land
17. GOODLEAS	whether farmers who leased their land to the oil-shale company got a good deal

Independent Variables

18. COUNTY	county of residence
19. HEARD	whether respondent had heard about the potential oil-shale development

20. AGE	year born
21. SEX	sex of respondent
22. EDUC	educational attainment in years
23. INCOME	family income
24. OCCUP	occupation of household head
25. OWNPROP	whether family owns property

Variables are described here to inform the reader in a manner consistent with the analysis. The authors were careful to distinguish the meanings of positive responses from negative responses in the interpretation.

8

Socioeconomic Impacts of Large-Scale Development Projects in the Western United States: Implications for Synthetic Fuels Commercialization

F. LARRY LEISTRITZ AND STEVE H. MURDOCK

As noted in other chapters of this work, a large number of synthetic-fuel facilities were proposed for development during the 1970s and early 1980s. A substantial percentage of these projects were planned for construction in rural areas with small populations and only limited infrastructure, often in the Great Plains or Rocky Mountain states. The tendency to site these facilities in rural areas relatively remote from major population centers arose both from economic factors, which often favored locating synfuel facilities close to the source of feedstock (such as coal or oil shale), and from environmental considerations, which discouraged the location of such plants near populous areas (Halstead et al., 1984). Because many synthetic-fuel projects were to be located in sparsely populated areas of the western states, it appeared likely that their development might lead to rapid economic, demographic, and social changes in nearby communities. As a result, considerable attention was focused on evaluating the nature of these effects (commonly termed *socioeconomic impacts*) of large-scale projects and on developing strategies for the alleviation of such impacts (often termed *impact management*).

Initial analyses of the socioeconomic effects of major projects created an image of "boomtown" growth as the likely consequence of large-scale industrial developments in rural areas (Kohrs, 1974; Gilmore and Duff, 1975). Major subsequent works have included those presenting theoretical bases for conducting impact assessments (Finsterbusch and Wolf, 1977; Murdock, 1979; Branch et al., 1984), proposing methodologies for assessing socioeconomic impacts (Finsterbusch and Wolf, 1981; Leistritz and Murdock, 1981), describing the general types of impacts likely to result from large-scale projects (Summers and Selvik, 1982; Murdock and Leistritz, 1979; Carley and Bustelo, 1984), providing retrospective analyses of the impacts of projects at multiple sites (England and Albrecht, 1984; Gilmore et al., 1982; Chalmers et al., 1982), evaluating the accuracy of impact assessments and assessment techniques (Murdock et al., 1984a, 1984b), and providing guides for ameliorating the socioeconomic impacts of large-scale developments (Weber and Howell, 1982; Halstead et al., 1984).

Within the last few years, an equally problematic form of impact associated with large-scale energy-development projects has also become apparent—the impact of project cancellation, abandonment, or closure. In a number of localities, unexpected decreases in the demand for a facility's product have abruptly changed an anticipated future of robust economic growth to one of mass unemployment, plummeting housing values, and fiscal shortfalls (Halstead et al., 1984; McGinnis and Schua, 1983). While these impacts have yet to receive the level of attention accorded to the impacts of growth, several recent works examine the effects of decline and potential mitigation measures (Hansen and Bentley, 1981; McKersie and Sengenberger, 1983; Halstead et al., 1984).

This chapter draws heavily on these previous works in first examining the nature of local socioeconomic impacts commonly associated with large-scale resource-development projects, such as synthetic-fuel plants, and then reviewing the nature and success of strategies adopted to cope with these effects. In addition, impact issues associated with facility closure or cancellation are briefly discussed. Because only a few commercial-scale synfuel projects actually reached the construction stage, evidence is drawn largely from other types of large-scale development projects in rural areas of the United States.

SOCIOECONOMIC IMPACTS OF RESOURCE DEVELOPMENT

In this section the state of knowledge related to each of several major categories of impacts—economic, demographic, public service and fiscal, and social—is examined. Although the categories shown are admittedly somewhat arbitrary, they have become widely recognized in the impact literature (Murdock and Leistritz, 1979; Leistritz and Murdock, 1981) and provide a useful basis for summarizing existing evidence concerning the socioeconomic impacts of large-scale developments. For each of these four types of impacts, the major issues and questions that have been the focus of attention in that area of impact analysis are summarized and available evidence concerning these impacts is presented.

Economic Impacts

The potential economic effects of large-scale projects have been a major focus of attention because the economic impacts are perceived to be among the most positive. Although numerous aspects of economic effects have been examined (Chase et al., 1983), the following three have received the most attention:

1. What proportions of project construction and operational jobs are likely to be filled by local workers, as compared to inmigrants, and what are the likely origins of inmigrating workers?
2. What is likely to be the magnitude of the secondary (indirect and induced) employment resulting from project development? What proportions of these jobs will be filled by local workers?
3. How will local businesses be affected by rapid growth resulting from a major project? For example, will development provide opportunities for expansion, or will local firms experience difficulty competing with new chain stores and in attracting and retaining quality workers?

An examination of the characteristics of work forces associated with large-scale development projects reveals substantial differences between construction work forces and permanent operations-and-maintenance work forces. For example, studies of workers at coal mine and power plant projects in the West and Southwest reveal that the construction work forces have been dominated by craftsmen with highly specialized skills who are geographically mobile in response to new job opportunities. Employment of a given craft at an individual project is temporary, training and apprenticeship periods for many crafts are relatively long, and labor unions play a major role in allocating workers to project sites (Wieland et al., 1979).

Because of these factors, many construction jobs are filled by relocating workers. A survey of workers at 14 energy-project construction sites in the western states indicated that relocating workers made up 60 percent of these work forces (Mountain West Research, 1975). Another survey of workers at 12 water-development projects in the western states revealed that nonlocal workers made up 53 percent of these work forces (Chalmers, 1977). However, a United States Army Corps of Engineers survey of workers at 55 mostly small construction projects indicated that only 31 percent of these workers were nonlocal (Dunning, 1981). The rate of local hiring appears to be affected by a number of factors, including the size of the local labor force, the status of other construction projects in the area, and the union referral system. Local (nonrelocating) workers are most prevalent in the less-skilled job categories, where more than half of the laborers are often local workers. Local hiring is also more frequent in the more populous site areas and for smaller projects.

High rates of local hiring can be achieved even for large projects when the local labor pool is substantial. A survey of construction workers at 6 Tennessee Valley Authority power plant projects revealed that 71 percent of these workers were local residents (DeVeney, 1977). Similarly, a survey of 13 nuclear power plant sites indicated that local workers made up at least half of the construction work forces in all cases and generally more than two-thirds. Substantial variability was noted in this study, however, with local worker percentages ranging from 50 to 86 percent (Malhotra and Manninen, 1980).

In part, the high rates of local recruitment achieved in areas with substantial local labor pools result from extensive daily commuting by construction craftsmen (Metz, 1982). For example, about one-third of the construction work force of the Great Plains Gasification Project in western North Dakota commuted daily from towns at least 60 miles away (Pearson, 1984).

The operations-and-maintenance work forces at many resource-development projects are dominated by heavy-equipment operators, craftsmen-technicians, and mechanics. Wages paid are generally higher than those in other rural-area jobs, and many firms appear to emphasize local hiring in order to build a stable work force (Leistritz et al., 1982). For example, a survey of workers at 14 power plants and coal mines in the northern Great Plains indicated that local workers made up 62 percent of these work forces (Wieland et al., 1979), while a survey of workers at 2 coal mines near Sheridan, Wyoming, indicated that about 60 percent of the work force had been recruited locally (Hooper and Branch, 1983). Similar results were reported from a survey at the Jim Bridger power plant in southwestern Wyoming (Browne, Bortz, and Coddington, Inc., 1981). However, substantial variations in the rates of local hiring have been found among various projects; rates generally are higher for smaller projects and those located in more populous areas. On-the-job training and internal promotions are often used to fill skilled positions, although a small nucleus of experienced workers sometimes is transferred between projects when a new facility begins operation.

Analyses of the origins of nonlocal workers show considerable variability among job types and also among project sites. For instance, a survey of workers at 14 construction sites indicated that 46 percent of the nonlocal workers came from within the state where the project was located, 16 percent came from adjacent states, and 38 percent from other states (Mountain West Research, 1975). The percentage of nonlocal workers originating within the state ranged from a high of 91 percent to a low of 11 percent, with lower percentages generally occurring in more sparsely populated states. Workers with highly specialized crafts, such as pipefitters, boilermakers, and millwrights, tended to be drawn from a relatively wide area, while more abundant crafts, such as carpenters and cement finishers, were more likely to be hired from within the state (Baker, 1977).

Similar worker origin patterns have been observed for operations-and-maintenance workers. Surveys at 4 coal mine and power plant sites in Wyoming revealed that 51 percent of the workers had been employed within the state immediately before taking their present job and 13 percent had been working in an adjacent state. A similar analysis for 8 energy projects in North Dakota indicated that 65 percent of the nonlocal workers had been employed within the state prior to being hired by the energy facility (Murdock and Leistritz, 1979).

The secondary employment effects of a large-scale project also may be quite substantial. The amount of secondary employment likely to result from development of a project is often estimated by an employment multiplier, which expresses the change in total project-related employment (primary plus secondary) as a multiple of the original change in project employment.

The secondary employment effects of large-scale projects have been difficult for researchers to measure with precision. Early impact projections (Murdock and Leistritz, 1979) often incorporated multipliers of about 1.6 during the construction phase and about 2.5 for the operational period (implying 0.6 and 1.5 secondary jobs, respectively, for each project job). Recent impact evaluations, however, indicate that construction-phase employment effects may be substantially smaller than previously believed. Based on retrospective analysis of impacts at 12 power plant construction sites, Gilmore et al. (1982) estimate construction-phase employment multipliers to be 1.2 to 1.3 for rural, sparsely populated areas and only 1.3 to 1.4 for areas with moderate population densities. In a similar retrospective analysis of 12 nuclear power plants, Pijawka and Chalmers (1983) estimate that the average local employment multiplier was 1.16 during construction and 1.23 during operation of the facilities. Fookes (1981) developed a similar estimate of the local employment multiplier (1.125) for the construction of the Huntly Power Project in New Zealand. Pijawka and Chalmers attributed the small multiplier values to substantial leakages of purchasing power resulting from (1) a high percentage of the construction workers commuting daily from nearby metropolitan areas and (2) the developers purchasing few goods and services within the impact area. Fookes cites both of these factors but also indicates that a major effect of the Huntly project was to increase productivity (sales per worker) within existing firms rather than to stimulate creation of new firms or new jobs within existing firms.[1]

Although numerous studies have addressed the origins of project construction and operations workers, much less attention has been given to the workers who fill the secondary jobs resulting from project development. Past impact assessments have frequently incorporated the assumption that most of these new jobs will be filled either by local residents or by spouses of the inmigrating project workers (Berkey et al., 1977; Denver Research Institute, 1979). A recent study, however, suggests that this assumption may not be tenable. A survey of local trade and service firms was conducted in nine cities in four western states that had experienced substantial levels of energy-related growth. (In several of these towns, the population had more than doubled since 1970.) The results indicated that 56 percent of the employees had moved to these counties since project development began. Of these recently inmigrating workers, only 18 percent had an immediate family member employed at one of the energy facilities. Overall, then, nearly 40 percent of the local trade and service sector workers in these communities were recent inmigrants and were not associated with relocating energy-project workers. Since each worker had an average of 1.7 associated family members, these inmigrating secondary workers constituted a significant source of local population growth (Halstead and Leistritz, 1983b).

The prospect of rapid growth associated with development of large-scale projects has led to several questions concerning the outlook for local businesses. Two particularly salient issues relate to the possibilities that (1) chain stores or other new firms with better access to managerial and capital resources may enter the local market and (2) local firms may experience rising labor costs and increased worker turnover as they are forced to compete with the new resource-development firms.

With respect to competition from new firms, recent studies provide conflicting results. For instance, a study of the Colorado oil-shale region indicates that chain stores have frequently moved into these communities and that their access to financing provides a competitive advantage (Denver Research Institute, 1979). On the other hand, a recent survey of six counties experiencing substantial energy-related growth in four western states suggested a very limited role for chain stores. Only 8.4 percent of the businesses surveyed in these counties were national or regional chains, and only one-third of these had been established subsequent to the advent of energy development in these counties. At the same time, more than half of the businesses surveyed had been established within the past ten years, and many of the remaining ones had expanded during this period. Thus, business expansion and new business opportunities in these counties were associated with the energy developments (Halstead and Leistritz, 1983b).

Local trade and service firms in rapid-growth areas have often been assumed to experience problems of increased worker turnover, difficulty in attracting quality workers, and pressure for substantial wage and salary increases to compete with the new development project. The results of the six-county survey referred to above, however, suggest that these problems may be less serious than previously believed. Only one-third of the businesses surveyed reported substantial increases in worker turnover since the advent of development activities. Similarly, less than 40 percent reported difficulty in attracting quality workers. Furthermore, while most businesses reported making "substantial" wage increases during the last five years, the percentage increases averaged only slightly more than the increase in the cost of living over the same period. Average hourly wages in 1983 for the businesses surveyed in these counties ranged from $4.00 to $6.40, suggesting that wage levels for local firms were not unduly inflated (Halstead and Leistritz, 1983b).

Available evidence related to local economic impacts suggests that both the direct and indirect economic benefits of large-scale projects may be more modest than is often anticipated and that, like other large-scale industrial developments in rural areas (Summers et al., 1976), the economic benefits of such projects may be overestimated prior to the initiation of such projects. At the same time, however, available evidence also indicates that the expected negative economic impacts on local businesses' market shares and labor forces may be less severe than anticipated. In sum, it seems that the local economic impacts of large-scale developments are neither inherently positive nor negative but rather are heavily dependent on the nature of the site area and of the project.

Demographic Impacts

Population growth has been one of the most publicized effects associated with large-scale developments in rural areas (Gilmore and Duff, 1975; Murdock et al., 1980). This growth has been characterized as having direct and often dramatic impacts on local communities; it is the source of many major social changes and a major impetus to many types of economic changes. Thus, the extent of growth that may be expected to result from large-scale developments, the distribution of such growth among local communities, and the effects that such growth is likely to have on the characteristics of the local population have been areas of paramount concern for impact researchers and policy makers (Murdock and Leistritz, 1979; Halstead et al., 1984).

Existing analyses have made it apparent that the level of local demographic impacts is a function of three general sets of factors: the characteristics of the project (such as size, location, skill requirements, and the like); the characteristics of the local areas (such as skill levels of local workers); and the characteristics of project-related inmigrants (such as their age, income, and preferences) (Murdock and Leistritz, 1979; Murdock et al., 1982a). Retrospective analyses (Gilmore et al., 1982; Chalmers et al., 1982) have established that it is the unique combination of these factors and the interaction of such characteristics that determine the magnitude and the nature of local demographic impacts. Despite considerable geographic diversity, however, generalizable information has been accumulating related to several key issues concerning demographic impacts. Of particular concern have been the answers to such questions as the following:

1. What levels of growth have actually occurred in developing areas?
2. What types of communities have experienced the most rapid growth?
3. What are the characteristics of inmigrants to developing areas, and how do their characteristics compare to those of longtime residents?

Several recent analyses have examined the levels of population growth associated with large-scale developments (Murdock et al., 1982a and 1982b; Gilmore et al., 1982). These analyses suggest that the population growth accompanying development of large-scale projects is likely to be substantial, but not as large as the boomtown literature has sometimes maintained. Such developments may, in fact, induce population stability rather than excessive growth. Growth or stability may not be areawide but may be quite localized. Growth is likely to slow once construction is complete and may display a pattern of decline when the project is ended. Thus, the growth related to large-scale resource developments may be more limited than is often believed. Other recent retrospective evaluations of the accuracy of projections of project-related growth (Murdock et al., 1982a and 1984a) also suggest, however, that the level of such growth is extremely difficult to predict.

The question of which types of communities have experienced the most rapid growth has only recently been evaluated (Murdock et al., 1982a and 1984b), but it appears that growth is likely to be concentrated in those areas with larger population bases. For example, one analysis of 84 impacted communities (Murdock et al., 1982a) found that although percentage increases in population were often largest in the smaller town, communities with initial populations under 1,000 gained an average of only 282 people for the ten years between 1970 and 1980, while communities of 1,000 to 2,499 gained an average of 1,290 people, and communities of over 2,500 population gained an average of 3,535 people per community during the ten-year period. Such analyses, plus the evidence from retrospective analyses (Chalmers et al., 1982; Gilmore et al., 1982; Murdock et al., 1984a), suggest that large-scale developments are unlikely to "save" the very smallest cities in rural areas.

The characteristics of inmigrating workers have been examined in numerous analyses (Mountain West Research, 1975; Wieland et al., 1979; Murdock and Leistritz, 1979; Murdock et al., 1980 and 1982a; Gilmore et al., 1982; Halstead et al., 1984). There is considerable agreement in the findings from these various studies and considerable agreement with studies of migrants in general. That is, migrants associated with large-scale developments appear not to be very different from other migrants. Comparisons of longtime residents to migrants reveal that migrants to such developments—like migrants in general (Ritchey, 1976)—are positively selected on the basis of age, education, and income. They are younger, better educated, and likely to have higher incomes than longtime residents (Murdock et al., 1980). They are likely to have smaller families, are more likely to live in mobile homes, and are more likely to obtain a higher proportion of the skilled and managerial jobs at a project than longtime residents (Mountain West Research, 1975; Gilmore et al., 1982).

Overall, then, the available evidence in regard to the demographic impacts of large-scale developments suggests that such impacts are unlikely to result in boomtown levels of growth but that they are difficult to predict and highly variable depending on the characteristics of the project, the local area's population, and the characteristics of new inmigrants. The characteristics of these inmigrants, however, tend to be similar to those of other migrants, and thus it appears likely that they have provided some resurgence of demographic vitality for many rural areas.

Public Service and Fiscal Impacts

The public service and fiscal impacts of resource developments have been major areas of concern in socioeconomic impact assessments (Denver Research Institute, 1979; Murphy and Williams, 1978). Changes in the availability and quality of public services and the revenues and costs associated with a major project are among the most visible and widely noted impacts of development (Summers et al., 1976; Lonsdale and Seyler, 1979) and have been a topic of much discussion in the impact literature (Gilmore and Duff, 1975; Halstead et al., 1984).

In these areas of impacts, the key questions have been ones such as the following:

1. Which service sectors are likely to be most and which are likely to be least impacted by large-scale developments?
2. To what extent are levels of service satisfaction likely to be impacted and which services are likely to be perceived as the most negatively impacted?
3. Do developments pay their own way in relation to public costs by generating sufficient revenue to meet the costs of increased public service demands?

Although the service sectors most severely impacted have varied from site to site depending on such factors as predevelopment service levels, the characteristics of the service area, and the service delivery system in the area, most available analyses indicate that impacts are likely to be most severe in relation to housing (Gilmore et al., 1982), recreation (Uhlman and Olson, 1984), and social services (Davenport and Davenport, III, 1979). In these areas it is evident that existing levels of services are often inadequate and that increased service demands often create crises in service delivery. For water, sewer, and transportation, prevailing evidence suggests that services are seldom impacted beyond manageable levels. For other community services, the results are more mixed. For example, educational and hospital services are often positively impacted by growth because baseline trends have led to declines in service needs and to decreased service capabilities (Schriner et al., 1976). In other cases, however, such services have been overly taxed, leading to significant service disruption (Davenport and Davenport, 1979). Such findings have clearly reinforced the fact that service impacts vary widely from one type of service to another and from area to area.

The question of impacts on service satisfaction also has been examined in several analyses (Murdock and Schriner, 1979; England and Albrecht, 1984; Albrecht et al., 1985). These analyses provide substantial consensus that dissatisfaction with services does increase during the construction and early operational phases of a large-scale development and that dissatisfaction levels are highest in regard to housing and recreation. They show some disagreement, however, concerning the persistence of such dissatisfaction; the analysis by England and Albrecht (1984) suggests that such dissatisfaction may persist, while the analyses by Murdock and Schriner (1979) and by Albrecht et al. (1985) suggest that service satisfaction levels may return to relatively high predevelopment levels after the project has reached a stable state of operation. These latter analyses also have shown that differences in levels of service satisfaction between new residents and indigenous residents do exist but that they are relatively small. Service satisfaction appears to vary more from structural changes in the area than from differences in individual characteristics.

These analyses also have established that many of the perceived problems with services in impacted areas stem from differences among residents' service-quality standards, preferences in relation to the form of service delivery, and their overall levels of service satisfaction (Murdock and Schriner, 1979). Finally, such analyses have clearly determined that the analysis of public services is really a multifaceted area of investigation requiring analyses of dimensions unique to each of the types of services being examined (Leistritz and Murdock, 1981).

A review of recent studies concerning fiscal effects of large-scale projects suggests that the answer to the question of whether projects pay their own way must be a qualified one. Specifically, it is critical to note that the fiscal implications of development depend on the time period being considered (construction versus operation), on the jurisdictions being considered, and on the ultimate fate of the project (will it be completed as planned?). Further, the tax status of some facilities has a substantial effect on their local fiscal implications. Finally, recent research clearly indicates that state and local governments generally have the capacity to substantially affect the fiscal implications of a development and ensure that the fiscal consequences will be satisfactory (Gilmore et al., 1982).

The timing of project-related revenues and expenditures poses problems for many local governments. The major timing problem arises from the fact that, although service demands and associated expenditures rise immediately during construction of a project, substantial revenues may not be received until operation of the facility begins (Figure 8–1). This situation is most common for local governments that depend primarily on property taxes because project facilities often are not taxed until completion and because construction workers often live in temporary housing with low taxable values. As a result, a critical revenue-expenditure squeeze may exist during the first several years of project construction (Gilmore et al., 1976; Murdock and Leistritz, 1979).

Fiscal problems of local government often are exacerbated when substantial project-related growth creates needs for investment to expand schools, water and sewer systems, and other public facilities. Local borrowing capacity often is limited by statute to a fixed percentage of the assessed valuation of a jurisdiction's tax base. In a rapid-growth situation, such limitations may severely limit a community's ability to expand its public facilities in a timely manner (Murdock and Leistritz, 1979; Halstead et al., 1984).

Another factor complicating effective planning for and management of growth associated with major projects is uncertainty regarding the extent of local growth that may actually be experienced. Uncertainty regarding whether

Figure 8–1
Employment, Tax Revenues, and Need for Public Services in an Area Affected by a Large-Scale Project

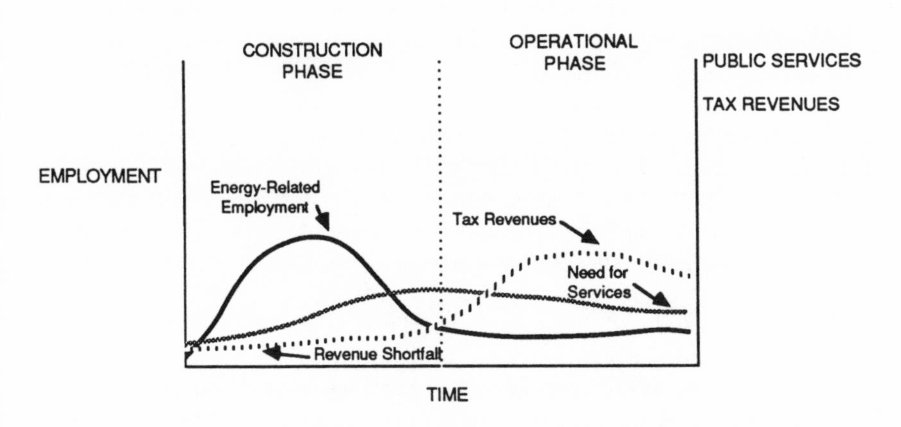

the project will actually take place, when it will begin, and whether it is possible that it will be developed only to be abandoned later as unfeasible—together with uncertainty regarding the potential impacts at the community level—can have substantial effects on community infrastructure financing decisions. Public officials and voters, cognizant of past boom and bust experiences associated with large projects, have often been reluctant to increase public debt if it is not clear that future increases in assessed valuation will be sufficient to repay the debt without substantial tax increases (Murdock and Leistritz, 1979). Recent closures and abandonments of a variety of resource facilities (for example, nuclear power plants, oil-shale projects, and mines) have led to an increased awareness of such issues and greater reluctance on the part of local officials and citizens to accept the risk associated with extensive borrowing (Halstead et al., 1984).

The distribution of project-related revenues and the tax structure of state and local areas also pose problems for fiscal management. In some cases, the facilities and resources that generate new revenues have been located in one county while the bulk of the project-related population is located in a different county or even a different state (Gilmore et al., 1982). In other situations, the majority of project-related revenues accrue to state and county governments while the bulk of project-related expenditures are borne by municipalities and school districts. In addition, in some states certain types of facilities (for example, publicly owned utilities) are tax exempt, and local jurisdictions may have to place heavy reliance on property taxes from non–project-related facilities. Thus, several analyses indicate that, in many cases, the major fiscal problems associated with large-scale projects are related to the distribution of revenues and expenditures or to the tax structure and not to the overall expenditure-revenue balance (Murray and Weber, 1982; Leistritz and Murdock, 1981).

Thus, the question "Does development pay its own way?" is not an easy one to resolve. Findings to date suggest that large, capital-intensive projects often generate revenues to state and local governments that tend to exceed total expenditures by these units over the life of the facility. However, local governments may experience substantial periods early in the project's life cycle in which revenues fall short of expenditures, and those local jurisdictions that are not able to directly tax the facility but host much of the associated population may experience chronic fiscal stress. As these characteristics of large-scale projects have become more widely understood, state and local officials and project managers have attempted to develop new institutional mechanisms, or modify existing ones, in order to alleviate the local fiscal problems associated with large projects (Halstead et al., 1984).

Thus, the existing evidence related to fiscal impacts suggests that the area is one where few overall generalizations can be drawn. Even more than in the other impact areas described above, the results in this area of analysis suggest that the impacts are areally specific. The circumstances surrounding the location and taxation status of the facility determine the magnitude and the nature of the impacts experienced. Although one might conclude, on the basis of existing evidence, that projects eventually pay for their most direct service costs, such a conclusion would likely have little generalizability to future projects occurring in different fiscal and geographic settings.

Social Impacts

The social impacts of large-scale developments often have been inappropriately defined to include nearly all other impacts not analyzed in the economic, demographic, and public service and fiscal impact areas. Thus, social impacts often have been seen as including psychological, historical, and many other dimensions that are not primarily sociological (Murdock and Leistritz, 1979). In addition, the assessment of such impacts often has been seen as a strategy to be used in gaining public acceptance of large-scale and unpopular projects (Branch et al., 1984). Even when appropriately defined to include the major social-structural and social-psychological dimensions basic to sociological analysis (Branch et al., 1984; Finsterbusch et al., 1983; Leistritz and Murdock, 1981), the focus of social impact analysis has been broad. Perhaps the most commonly raised questions concerning such impacts, however, have focused on the following issues:

1. Do large-scale projects alter the social interaction patterns and social-structural composition of rural communities?
2. Do such projects lead to major disruptions in social control mechanisms in rural

areas and thus result in increased rates of crime, delinquency, marital dissolution, and the like?

3. Which groups are most positively and which are most negatively impacted by such projects (for example, the elderly, the poor, the young)?
4. What levels of social-psychological stress are placed on persons living in the siting areas of large-scale projects, and, if stress is induced, does it have temporary or permanent effects on area residents?
5. Overall, do rural residents perceive large-scale projects as having had positive or negative impacts on their communities, and which aspects do they believe have been most positively and negatively impacted?

Some of the early analyses of social impacts (Kohrs, 1974; Gilmore and Duff, 1975) suggested that large-scale projects would permanently alter the interaction patterns and the social structures of rural communities, making interaction patterns less informal and leading to alterations in the sustenance and occupational bases of rural areas. Several analyses have refuted such hypotheses, however. Murdock and Schriner (1978) analyzed nine western communities and found that although there were dissimilarities between the occupational structures of impacted and nonimpacted areas during construction periods, the occupational structures of postimpact areas were generally not dissimilar from those in rural communities that had not experienced such projects. In a more recent analysis, England and Albrecht (1984) found little evidence to support the contention that the social structures of rural communities had been permanently altered as a result of such projects. Other analyses (Murdock et al., 1981; Gilmore et al., 1982; Chalmers et al., 1982), although with some qualifications, have come to similar conclusions.

The social disruption hypothesis is among the oldest in the social impact literature. The early work of Kohrs (1974) and Gilmore and Duff (1975) suggested that western (and other) boomtowns experienced disproportionate increases in crime, child and spouse abuse, divorce, and other forms of social disruption during the development of large-scale projects. Additional support for this hypothesis was found in the area of mental health (Davenport and Davenport, 1979). Wilkinson et al. (1982), however, have argued that when appropriate population bases are employed, it does not appear that the rates of such behaviors have increased, but only the absolute numbers of such incidents. Freudenburg et al. (1982), Albrecht (1982), and others (for example, Lantz and McKeown, 1977) suggest findings that contradict the Wilkinson et al. hypothesis. However, recent analyses (including those by England and Albrecht, 1984; Murdock et al., 1981; Wilkinson et al., 1983 and 1984) have tended to support the original findings of Wilkinson et al. Although there remains substantial disagreement about the extent to which disruption does occur (Murdock and Leistritz, 1982), it appears that levels of social disruption often have been lower than those anticipated in the boomtown literature.

Much of the early impact literature also suggested that large-scale projects might be particularly negative for certain groups who lacked the resources to manage the economic and other impacts of large-scale projects. The poor, youth, women, and particularly the elderly were seen as likely victims of such projects (Murdock and Leistritz, 1979). Recent analyses by Freudenburg (1984) and by Albrecht et al. (1985), among others, suggest that such negative impacts have been restricted to only a few groups. In general, the elderly have not been negatively impacted (Gilmore et al., 1982), and minorities and women also have not suffered disproportionately. In fact, available evidence (Freudenburg, 1984) suggests that it is the youth who have experienced the greatest stress and who have been most negatively impacted. Faced with the dual adjustment to adolescence and to the conditions of rapid community growth and lacking the established bases of social support of adults, they tend to experience substantial difficulty adjusting to rapid population growth. For nearly all other groups, however, there is little evidence of large-scale or lasting disruptions due to rapid growth.

The question of whether large-scale and potentially dangerous projects induce stress has been a particularly controversial area of analysis (Freudenburg and Jones, 1984; Murdock et al., 1983; Krannich et al., 1984) due to the recent ruling by the Supreme Court in regard to the accident at Three Mile Island (U.S. Supreme Court, 1983). In fact, there is a relatively substantial body of literature relating to stress and other psychological effects of major accidents (Fritz and Marks, 1954; Dynes, 1974; Sills et al., 1982). The existing evidence regarding these effects is very mixed, however. Freudenburg (1982 and 1984) and others (Weisz, 1979; Lantz and McKeown, 1977) have found extensive levels of stress, at least for particular groups (such as youth), but others have not found such effects (Krannich et al., 1984). Still other analyses suggest that such effects tend to dissipate rapidly and to largely disappear after the peak impact period has passed (Rossi et al., 1978). Additional research on the intensity and duration of the stress-related effects of large-scale projects is clearly necessary.

The perceptions of residents of rural areas concerning the overall impacts of large-scale developments have received considerable attention (Gilmore and Duff, 1975; Murdock and Leistritz, 1979; Thompson and Blevins, 1983). Early work (Gilmore and Duff, 1975) suggested that residents, though initially in favor of developments, tended to view such projects as negative once they were initiated. Later work by Murdock and Leistritz suggested that a majority of residents remained in favor of developments throughout the development period. The level of support for development tended to be cyclical, however, being highest prior to the construction of the project, decreasing during the construction stage of the project, and then increasing to high levels of support after the project reached the operational stage (Murdock and Leistritz, 1979). Recent analyses by Thompson and Blevins (1983), England and Albrecht (1984), and Albrecht et al. (1985) have provided substantial support for the Murdock and Leistritz (1979) hypothesis. It appears, then, that large-scale projects are seen as positive overall.[2]

Such analyses (particularly Albrecht et al., 1985, and England and Albrecht, 1984) also suggest that there is substantial variation in the evaluations of different types of impacts. These and similar analyses (Murdock and Schriner, 1978; Gilmore et al., 1982) indicate that residents perceive economic impacts to have been the most positive and social and public service impacts to have been the most negative.

Overall, then, available evidence on the social impacts of large-scale developments suggests that these impacts seldom resemble the boomtown syndrome often hypothesized in the early impact literature. Such impacts can be quite negative, but evidence suggests that they have tended to have both positive and negative impacts, with the most negative impacts being restricted to selected groups and areas. Thus, analysis in this area has served to replace the boomtown stereotype with increasingly sound empirical evidence.

Impacts of Cancellation or Closure

In the early 1980s, several communities where synfuel projects were proposed for development or were even under construction experienced a different type of impact —the cancellation or abandonment of a major project. Two noteworthy examples of this type of impact are the cancellation of the Colony Oil Shale project in western Colorado (after an investment of over $2 million) and the substantial downsizing of the Alsands tar sands project in northern Alberta. In these and other cases, the affected communities have experienced increased unemployment, reduced retail sales, outmigration of workers and their families, and a declining tax base. Effects on local governments have been especially severe if these units had borrowed heavily to develop expanded infrastructure to meet the needs of the anticipated project-related population growth.

In addition to these economic effects, facility abandonment may seriously impair the quality of life within the community. Individuals experiencing job loss may tend to withdraw from community life, leading to reduced participation in such social organizations as churches and volunteer groups (Hansen and Bentley, 1981). This withdrawal tends to dampen community vitality, weaken the social fabric, and contribute to a dreary climate that encourages further outmigration (McKersie and Sengenberger, 1983). Thus, the impacts of facility cancellation, abandonment, or closure are increasingly becoming recognized as an area requiring greater attention.

IMPACT MANAGEMENT—CONTROLLING THE EFFECTS OF DEVELOPMENT

As noted previously, development of large-scale resource development projects in remote areas can result in periods during which housing, public services, and private-sector retail and service capabilities are less than adequate to meet the needs of the expanded local population. Such situations have sometimes led to dissatisfaction on the part of both long-term residents and newcomers and to a general feeling that the local quality of life has been degraded. These feelings, in turn, have sometimes contributed to increased rates of labor turnover and absenteeism, decreased worker productivity, and increased development costs (Metz, 1983). Thus, project proponents, as well as the affected communities and states, have taken steps to ensure that the demands imposed by new development can be accommodated without undue strain.

Impact mitigation was the first term widely used to describe such efforts to alter the effects of development projects. However, mitigation has often been viewed in the relatively narrow context of merely reducing or eliminating negative impacts. As a consequence, several authors have suggested that the term *impact management*, connoting a more comprehensive approach encompassing measures to enhance the project's local benefits and to provide various forms of compensation to local interests, as well as actions to reduce or eliminate negative effects, would be more appropriate (Gilmore et al., 1982; Halstead et al., 1984; Leistritz and Ekstrom, 1986).

The types of impact management measures that have been undertaken or proposed in connection with synthetic-fuel plants or other large-scale energy facilities in the western United States fall into two general categories: (1) measures to minimize demands on local systems (economic, governmental, and social) and (2) measures to enhance the capacity of local systems to cope with change. The extent to which these various actions may be applicable to a specific project will depend mainly on the nature and location of the project and the institutional setting within which it occurs. In the remainder of this section, the specific types of measures within each category are examined, factors affecting their applicability are discussed, and pragmatic lessons from their use in connection with western energy projects are reviewed.

Minimizing Demands on Local Systems

Mitigation measures that reduce the demands imposed on local systems have a special appeal. These measures concentrate on avoiding the problems associated with overtaxed local economic, governmental, and social systems by reducing the number of inmigrants associated with a project. Such measures can thus be characterized as an attempt to avoid impact problems rather than to solve them after they occur. Approaches to achieving this goal fall into two general categories: (1) alterations in facility design or construction schedule to reduce peak labor requirements and (2) work force policies geared to reducing the proportion of workers that will migrate to the site area (Table 8–1).

There are three major alterations in facility design or construction schedule

Table 8–1
Measures to Reduce Inmigration

Option	Method	Advantages	Disadvantages/ Limitations
1. Alterations in facility design for construction schedule	1. Lengthening construction period	- reduces peak work force requirements - reduces number of relocating workers	- increases project cost
	2. Off-site component fabrication	- increases construction efficiency - eases needs for craftsmen and engineers on-site	- may be limited by union-employer contractual agreements and/or capability to transport large components
	3. Scheduling of multiple units	- reduces peak work force requirements	- is applicable only to projects with multiple units or when several projects are planned for the same area
2. Reducing the percentage of immigrating project workers and families	1. Increasing local hiring a. local hiring preference b. training programs	- increases percentage of economic benefits accruing to local residents - enhances stability of work force - may reduce projects' competition with local employers - is popular with local residents - increases number of workers hired locally	- may violate union-labor agreements - may be viewed as discriminatory - increases competition for local labor with area businesses - succes depends on current employment situation in area - may not increase local firing due to union-labor contractual agreements
	2. Increasing commuting a. measures to reduce travel costs b. provision of temporary housing	- increases ease of labor force recruitment - increases productivity - reduces number of relocating workers	- requires careful initiation, lead-time consideration - may aggravate local traffic problems - induces payroll leakages from local community - may lead to higher turnover

that could reduce the demands imposed on the site area: (1) lengthening the project construction period, (2) fabricating some facility components at off-site locations, and (3) staggering unit construction schedules for multiple-unit facilities (Table 8–1). The economic forces and geological imperatives that affect the siting and development of resource projects, however, seem to limit the applicability of these measures in many cases.

Given the project site and construction parameters, the principal method for lessening local impacts is to reduce the proportion of inmigrating workers. This strategy has two major options. The first is to increase the number of local workers hired through local hiring-preference or training programs. Increased local hiring will reduce migration into the area and will enhance local economic benefits, thereby increasing community acceptance of the project. Local hiring may be limited, however, by a number of factors (Table 8–1). These factors appear to be particularly relevant during the construction phase of many major projects; thus, local hiring is generally seen as having greater potential with respect to the project's operation-phase labor requirements than for the construction-phase labor force. The relatively short duration of the construction period and the key role of the union referral system in project staffing are the major factors that limit the potential for substantially affecting local hiring rates during the construction period. In order to reduce construction-phase inmigration and associated demands, therefore, it may be necessary to examine measures aimed at encouraging commuting as an alternative to relocation.

Measures to encourage commuting are intended to reduce workers' travel costs or to provide convenient accommodations for those who commute weekly. Measures to reduce workers' travel costs may include direct provision of transportation or provision of travel allowances. Temporary housing could be provided on-site to encourage workers to commute on a weekly basis.

Temporary housing was used extensively during construction of the Great Plains Coal Gasification Project near Beulah, North Dakota. During peak construction periods, several hundred workers were accommodated in single-status housing provided by the developers (Halstead and Leistritz, 1983a).

Increasing commuting may generate additional problems, however. First, traffic problems on local roads may be aggravated. An increase in multiple ridership, perhaps through increased use of buses and vans, is a means of alleviating this problem. Second, increased commuting may be seen by locals as causing project benefits to leave the area. Finally, high levels of commuting may lead to higher worker turnover rates (Halstead et al., 1984).

Enhancing the Capacity of Local Systems

Even though the level of inmigration associated with project development may be reduced, a large-scale facility will generate significant population growth and increases in the demands on local systems. It is necessary, therefore, to consider impact management measures that enhance the capability of local systems to cope with change. The literature suggests that such measures fall into five general categories: (1) local planning assistance, (2) provision by the developer of housing and infrastructure, (3) financial assistance to the local public sector, particularly front-end financing, (4) stimulation of housing development and business expansion by the local private sector, and (5) enhancement of local protection capabilities (Halstead et al., 1984).

One of the most important steps in successful impact management is initiating advanced planning at the local level. This is a common problem for small communities, which seldom have the capabilities to manage rapid growth. An important consideration in many western energy-development areas was the need to initiate the local planning phase well before project development began. Funding and technical assistance provided either by state government or by the developer directly have been instrumental in facilitating the planning process in many of these areas.

The most obvious problem encountered by impacted communities often is inadequate housing for the new population. The relocation of a large portion of the project work force will increase area housing demand. In the absence of adequate planning, this increase may result in inflated housing prices, rental fees, property values, and property taxes.

Since it is in the best interest of the developer to assure an adequate supply of housing near the project, measures have often been taken to increase the housing stock. These include construction of a self-contained community, development of permanent housing or mobile home parks, revitalization of existing housing, or provision of temporary housing. While energy-development firms have frequently provided housing and related infrastructure, most efforts by developers have been aimed at stimulating private housing developers (Halstead et al., 1984).

These efforts often take the form of assembling land suitable for development or guaranteeing purchase or occupancy of new housing units. Because uncertainty often inhibits local developers from initiating new housing construction, such guarantees will often be effective in stimulating development. Similarly, uncertainty concerning the future of the project and the magnitude of local growth may retard local business expansion. If such expansion is perceived as desirable, the resource developer can stimulate it by establishing long-term contracts for supplies or services with selected local businesses. With such contracts in hand, local businesses typically find financing for expansion much easier to obtain.

When rapid development occurs, local governments typically face a cost-revenue squeeze. Substantial capital costs often must be incurred to provide for necessary expansion of service capacity at a time well before significant development-related revenues are received. In such cases, it may be necessary for local governments to rely on grants from senior governments, prepayment of sales or property taxes by the developer, borrowing, or other front-end financing mechanisms.

In areas of the western United States that have experienced substantial levels of energy-resource development, several sources of financial assistance have been utilized by affected local governments. First, some state governments have provided substantial levels of financial aid. Several states, including Colorado, Montana, North Dakota, and Wyoming, have earmarked a portion of their mineral severance taxes to aid affected communities. In some states, state grants and loans have been the primary source of financing for community facilities, while in others, industry also has been expected to provide direct assistance to communities (Halstead et al., 1984). Financing from industry has taken a variety of forms, including prepayment of taxes, loans, loan guarantees and outright grants to local jurisdictions. Finally, the federal government has provided a significant amount of financial resources for improving public service infrastructure in impacted communities. Most of these funds, however, have been made available through categorical grant programs that are not specifically targeted to the needs of rapid-growth areas. A considerable level of local expertise, as well as persistence, has been required for communities to benefit from these programs, and hence, these programs generally are not regarded as a dependable source of financial assistance (Halstead et al., 1984).

It is perhaps too soon to objectively assess the success of impact management programs in altering the effects of energy-resource development in the western United States. However, several generalizations can be drawn based on a review of the literature to date. The first of these is simply that at least some of the impacts of large-scale development do appear to be manageable. The massive relocation of project-related workers and their families, for example, can be reduced substantially through a combination of work force management measures. Similarly, many potential shortfalls in public services and infrastructure can be forestalled by a combination of timely planning and the availability of adequate financial assistance (Halstead et al., 1984).

A second generalization is that anticipatory impact assessments often have been somewhat inadequate as a guide for impact management. Changes in project timetables and labor force requirements have often caused both the timing and the magnitude of socioeconomic change to differ markedly from the indications of the anticipatory assessment. A common response to this problem has been to incorporate socioeconomic monitoring into the impact management process (Leistritz and Chase, 1982). An effective monitoring system can provide timely information, which enables decision makers to assess community needs and, if necessary, revise mitigation plans to meet changing circumstances.

A final generalization is that the need to recognize the degree of uncertainty associated with development of large-scale resource projects is hard to overemphasize. Recent abandonments of a number of major synfuel projects, such as the Colony oil-shale project in Colorado, have made local officials and residents of energy-rich areas increasingly aware of the risks associated with such projects. The issue of how these risks are to be shared between the local community, higher levels of government, and the developer will likely become a major one when future projects are proposed and impact management plans developed.

CONCLUSIONS AND IMPLICATIONS

When massive development of synfuel projects was proposed, some observers suggested that the economic, demographic, and social effects of such development would devastate nearby communities and that concerns regarding such impacts would pose a major obstacle to the deployment of synfuels technology in rural areas. Subsequent research has provided a somewhat more objective assessment of the nature and extent of socioeconomic impacts associated with large energy projects, and experience gained in attempting to mitigate impacts of large-scale projects suggests that if planning is timely and adequate financial resources are allocated to the task, many of the impacts of such projects are manageable. One lesson that may be drawn from the experience of the past decade, then, is that a future deployment of synfuels technologies could be accomplished without creating unacceptable levels of impacts for nearby communities.

NOTES

1. While these analyses point out the importance of considering linkages when estimating secondary employment effects, it also should be noted that their multiplier estimates are influenced by their definition of the relevant impact area. Thus, Gilmore et al. (1982) estimate secondary employment for the site county only. Fookes's estimate is for the site community (Huntly) only, while Pijawka and Chalmers's estimates are for either the site county or a subcounty area. Had all of the impact areas been defined more broadly in these studies (to include the cities where many of the workers resided), the multiplier estimates might have been considerably larger.
2. A recent analysis by Freudenburg and Baxter (1984), however, suggests that perceptions of large-scale nuclear projects may have been permanently altered by the accident at Three Mile Island such that residents remain skeptical about the benefits of such projects.

REFERENCES

Albrecht, D. E.; Murdock, S. H.; Halstead, J.; Leistritz, F. L.; and Albrecht, S. 1985. "The Impacts of Large-Scale Developments on Rural Communities in the Western United States." In *Research in Rural Sociology and Development*, ed. by H. K. Schwarzweller and F. A. Fear. Greenwich, Conn.: JAI Press, pp. 109–123.

Albrecht, S. L. 1982. "Commentary." *Pacific Sociological Review* 25(3):297–306.

Baker, J. S. 1977. *Labor Allocation in Western Energy Development*. Monograph no. 5. Salt Lake City: University of Utah, Human Resources Institute.

Berkey, E.; Carpenter, N. G.; Metz, W. C.; Meyers, D. W.; Portes, D. R.; Singley, J. E.; and Travis, R. K. 1977. *Social Impact Assessment, Monitoring, and Management by the Electric Energy Industry*. Pittsburgh, Pa.: Energy Impact Associates.

Branch, K.; Hooper, D. A.; Thompson, J.; and Creighton, J. 1984. *Guide to Social Assessment: A Framework for Assessing Social Change*. Boulder, Colo.: Westview Press.

Browne, Bortz, and Coddington, Inc. 1981. *A Retrospective Analysis of the Jim Bridger Complex Socioeconomic Monitoring Report*. Denver: Browne, Bortz, and Coddington, Inc.

Carley, M. J., and Bustelo, E. S. 1984. *Social Impact Assessment and Monitoring: A Guide to the Literature*. Boulder, Colo.: Westview Press.

Chalmers, J. A. 1977. *Construction Worker Survey*. Denver: U.S. Bureau of Reclamation.

Chalmers, J. A.; Pijawka, D.; Branch, K.; Bergman, P.; Flynn, J.; and Flynn, C. 1982. *Socioeconomic Impacts of Nuclear Generating Stations*. NUREG/CR–2750. Washington, D.C.: Government Printing Office, U.S. Nuclear Regulatory Commission.

Chase, R. A.; Leistritz, F. L.; and Halstead, J. M. 1983."Assessing the Economic and Fiscal Effects of Repository Development." In *Nuclear Waste: Socioeconomic Dimensions of Long-Term Storage*, ed. by S. H. Murdock, F. L. Leistritz, and R. R. Hamm. Boulder, Colo.: Westview Press, pp. 119–134.

Davenport, J. A., and Davenport, J., III, eds. 1979. *Boom Towns and Human Services*. Laramie: University of Wyoming, Department of Social Work.

Denver Research Institute. 1979. *Socioeconomic Impacts of Western Energy Resource Development*. Washington, D.C.: Council on Environmental Quality.

DeVeney, G. R. 1977. *Construction Employee Monitoring*. Knoxville: Tennessee Valley Authority.

Dunning, C. M. 1981. *Report of Survey of Corps of Engineers Construction*. Institute for Water Resources Research Report 81-R05. Fort Belvoir, Va.: U.S. Army Corps of Engineers.

Dynes, R. R. 1974. *Organized Behavior in Disaster*. Lexington, Mass.: D. C. Heath and Co.

England, J. L., and Albrecht, S. L. 1984."Boomtowns and Social Disruption." *Rural Sociology* 49(2):230–46.

Finsterbusch, K., and Wolf, C. P., eds. 1977. *Methodology of Social Impact Assessment*. 1st ed. Stroudsburg, Pa.: Dowden, Hutchinson, and Ross.

———. 1981. *Methodology of Social Impact Assessment*. 2d ed. Stroudsburg, Pa.: Dowden, Hutchinson, and Ross.

Finsterbusch, K.; Llewellyn, L. G.; and Wolf, C. P., eds. 1983. *Social Impact Assessment Methods*. Beverly Hills, Calif.: Sage Publications.

Fookes, T. W. 1981. *Expectations and Related Findings, 1973–81*. Hamilton, New Zealand: University of Waikato.

Fritz, C. E., and Marks, E. S. 1954. "The NORC Studies of Human Behavior in Disaster." *The Journal of Social Issues* 10(3):26–41.

Freudenburg, W. R. 1982. "Balance and Bias in Boomtown Research." *Pacific Sociological Review* 25(3):323–338.

———. 1984. "Differential Impacts of Rapid Community Growth." *American Sociological Review* 49(5):697–705.

Freudenburg, W. R., and Baxter, R. K. 1984. "Host Community Attitudes toward Nuclear Power Plants: A Reassessment." *Social Science Quarterly* 65(4):1129–36.

Freudenburg, W. R., and Jones, T. R. 1984. "Does an Unpopular Facility Cause Stress? A Test of the Supreme Court Hypothesis." Scientific Paper no. 6873, Research Project 0478. Pullman: Washington State University, Agricultural Research Center.

Freudenburg, W. R.; Bacigalupi, L.; and Young, C. Winter 1982. "Mental Health Consequences of Rapid Growth: A Report from the Longitudinal Study of Boom Town Mental Health Impacts." *Journal of Health and Human Resources Administration* 4:334–52.

Gilmore, J. S., and Duff, M. K. 1975. *Boom Town Growth Management: A Case Study of Rock Springs–Green River, Wyoming*. Boulder, Colo.: Westview Press.

Gilmore, J. S.; Moore, K. D.; and Hammond, D. M. 1976. *Synthesis and Evaluation of Initial Methodologies for Assessing Socioeconomic and Secondary Environmental Impacts of Western Energy Resource Development*. WP no. 2. Denver: Denver Research Institute.

Gilmore, J. S.; Hammond, D. M.; Moore, K. D.; Johnson, J.; and Coddington, D. C. 1982. *Socioeconomic Impacts of Power Plants*. Report prepared for Electric Power Research Institute. Denver Research Institute.

Halstead, J. M., and Leistritz, F. L. 1983a. *Impacts of Energy Development on Mercer County, North Dakota*. Ag. Econ. Rpt. no. 170. Fargo: North Dakota Agricultural Experiment Station.

———. 1983b. *Impacts of Energy Development on Secondary Labor Markets*. Ag. Econ. Rpt. no. 178. Fargo: North Dakota Agricultural Experiment Station.

Halstead, J. M.; Chase, R. A.; Murdock, S. H.; and Leistritz, F. L. 1984. *Socioeconomic Impact Management: Design and Implementation*. Boulder, Colo.: Westview Press.

Hansen, G. B., and Bentley, M. T. 1981. *Problems and Solutions in a Plant Shutdown*. Logan: Utah State University.

Hooper, J. E., and Branch, K. M. 1983. *Big Horn and Decker Mine Worker Survey Report*. Billings, Mont.: Mountain West Research-North.

Kohrs, E. 1974. *Social Consequences of Boom Growth in Wyoming*. Casper: Central Wyoming Counseling Center.

Krannich, R. S.; Golesorkhi, B.; and Greider, T. 1984. "Rapid Growth and Personal Stress: An Assessment of Stress Levels, Network Ties and Social Support in Energy-Impacted Communities." Paper presented at the annual meetings of the Rural Sociological Society, College Station, Texas.

Lantz, A., and McKeown, R. 1977. *Rapid Growth and the Impact on Quality of Life in Rural Communities: A Case Study*. Denver: Denver Research Institute.

Leistritz, F. L., and Chase, R. A. 1982. "Socioeconomic Impact Monitoring Systems: A Review and Evaluation." *Journal of Environmental Management* 15:333–49.

Leistritz, F. L., and Ekstrom, B. L. 1986. *Social Impact Assessment and Management: An Annotated Bibliography*. New York: Garland Publishing.

Leistritz, F. L., and Murdock, S. H. 1981. *The Socioeconomic Impact of Resource Development: Methods for Assessment*. Boulder, Colo.: Westview Press.

Leistritz, F. L.; Murdock, S. H.; and Leholm, A. G. 1982. "Local Economic Changes Associated with Rapid Growth." In *Coping with Rapid Growth in Rural Communities*, ed. by B. A. Weber and R. E. Howell. Boulder, Colo.: Westview Press, pp. 25–62.

Lonsdale, R. E., and Seyler, H. L., eds. 1979. *Nonmetropolitan Industrialization*. New York: John Wiley and Sons.

McGinnis, K., and Schua, D. 1983. "The Downside of Nuclear Plant Construction." Paper presented at Northwest Regional Economics Conference, 5–7 May, at Bellingham, Washington.

McKersie, B., and Sengenberger, W. 1983. *Job Losses in Major Industries: Manpower Strategy Responses*. Paris: Organization for Economic Cooperation and Development.

Malhotra, S., and Manninen, D. 1980. *Migration and Residential Location of Workers at Nuclear Power Plant Construction Sites*. Vols. 1 and 2. Seattle: Battelle Human Affairs Research Center.

Metz, W. C. 1980. "The Mitigation of Socioeconomic Impacts by Electric Utilities." *Public Utilities Fortnightly* 106:34–42.

———. 1982. "Energy Industry Involvement in Worker Transportation." *Transportation Quarterly* 36(4):563–84.

———. 1983. "Industry Initiatives in Impact Mitigation." In *Proceedings of the Alaska Symposium on the Social, Economic, and Cultural Impacts of Natural Resource Development*. Fairbanks: University of Alaska, pp. 239–251.

Mountain West Research, Inc. 1975. *Construction Worker Profile*. Washington, D.C.: Old West Regional Commission.

Murdock, S. H. 1979. "The Potential Role of the Ecological Framework in Impact Analysis." *Rural Sociology* 44(3):543–65.

Murdock, S. H., and Leistritz, F. L. 1979. *Energy Development in the Western United States: Impact on Rural Areas*. New York: Praeger Publishers.

———. 1982. "Commentary." *Pacific Sociological Review* 25(3):357–66.

Murdock, S. H., and Schriner, E. C. 1978. "Structural and Distributional Factors in Community Development." *Rural Sociology* 43(3):426–49.

———. 1979. "Community Service Satisfaction and Stages of Community Development: An Examination of Evidence from Impacted Communities." *Journal of the Community Development Society* 10(1):109–24.

Murdock, S. H.; Leistritz, F. L.; and Schriner, E. C. 1980. "Migration and Energy Developments: Implications for Rural Areas in the Great Plains." In *New Directions in Urban-Rural Migration*, ed. by D. Brown and J. Wardwell. New York: Academic Press, pp. 267–290.

Murdock, S. H.; deMontel, J.; Leistritz, F. L.; Hopkins, P.; and Hamm, R. R. 1981. *An Analysis of the Construction Impacts of Coal Development in Rural Texas: The Case of Fayette County, Texas*. Rpt. no. 81–2. College Station: Texas Agricultural Experiment Station.

Murdock, S. H.; Leistritz, F. L.; and Schriner, E. C. 1982a. "The Demographic Impacts of Rapid Economic Development." In *Coping with Rapid Growth in Rural Communities*, ed. by B. A. Weber and R. E. Howell. Boulder, Colo.: Westview Press; pp. 63–96.

Murdock, S. H.; Leistritz, F. L.; Hamm, R. R.; and Hwang, S. S. 1982b. "An Assessment of Socioeconomic Assessments: Utility, Accuracy, and Policy Considerations." *Environmental Impact Assessment Review* 3(4):333–50.

Murdock, S. H.; Leistritz, F. L.; and Hamm, R. R. 1983. *Nuclear Waste: Socioeconomic Dimensions of Long-Term Storage*. Boulder, Colo.: Westview Press.

Murdock, S. H.; Leistritz, F. L.; Hamm, R. R.; and Hwang, S. S. 1984a. "An Assessment of the Accuracy and Utility of Socioeconomic Impact Assessments." In *Paradoxes of Western Energy Development*, ed. by C. M. McKell, D. G. Browne, E. C. Cruze, W. R. Freudenburg, R. L. Perrine, and F. Roach. Boulder, Colo.: Westview Press, pp. 265–296.

Murdock, S. H.; Leistritz, F. L.; Hamm, R. R.; Hwang, S. S.; and Parpia, B. 1984b. "An Assessment of the Accuracy of a Regional Economic-Demographic Projection Model." *Demography* 21(3):383–404.

Murphy and Williams, Consultants. 1978. *Socioeconomic Impact Assessment: A Methodology Applied to Synthetic Fuels*. Washington, D.C.: U.S. Department of Energy.

Murray, J. A., and Weber, B. A. 1982. "The Impacts of Rapid Growth on the Provision and Financing of Local Public Services." In *Coping with Rapid Growth in Rural Communities*, edited by B. A. Weber and R. E. Howell. Boulder, Colo.: Westview Press, pp. 97–114.

Pearson, C. 1984. *ITAT Construction Work Force Report*. Bismarck, N.Dak.: Inter-Industry Technical Assistance Team (Basin Electric Power Corp.).

Pijawka, D., and Chalmers, J. A. 1983. "Impacts of Nuclear Generating Plants on Local Areas." *Economic Geography* 59(1):66–80.

Ritchey, P. N. 1976. "Explanations of Migration." In *The Annual Review of Sociology II*. Palo Alto, Calif.: Annual Review, Inc., pp. 363–404.

Rossi, P. H.; Wright, J. D.; Wright, S. R.; and Weber-Burdin, E. 1978. "Are There Long-Term Effects of American Natural Disasters?" *Mass Emergencies* 3:117–32.

Schriner, E. C.; Query, J.; McDonald, T.; and Keogh, F. 1976. *An Assessment of the Social Impacts Associated with a Coal Gasification Complex Proposed for Dunn County, North Dakota*. Fargo: North Dakota State University.

Sills, D. L.; Wolf, C. P.; and Shelanski, V. B., eds. 1982. *Accident at Three Mile Island: The Human Dimensions*. Boulder, Colo.: Westview Press.

Summers, G. F., and Selvik, A., eds. 1982. *Energy Resource Communities*. Madison, Wis.: MJM Publishing Co.

Summers, G. F.; Evans, S. D.; Clemente, F.; Beck, E. M.; and Minkoff, J. 1976. *Industrial Invasion of Nonmetropolitan America: A Quarter Century of Experience*. New York: Praeger Publishers.

Thompson, J. G., and Blevins, A. L. 1983. "Attitudes toward Energy Development in the Northern Great Plains." *Rural Sociology* 48(1):148–58.

Uhlman, J. M., and Olson, J. K. 1984. *Planning for Rural Human Services*. Denver: Office of Human Development Services, Department of Health and Human Services.

U.S. Supreme Court. 1983. *Metropolitan Edison Co. vs. People Against Nuclear Power (PANE)*. 103 S. Ct. 1556. Washington, D.C.

Weber, B. A., and Howell, R. E., eds. 1982. *Coping with Rapid Growth in Rural Communities*. Boulder, Colo.: Westview Press.

Weisz, R. 1979. "Stress and Mental Health in a Boom Town." In *Boom Towns and Human Services*, ed. by J. A. Davenport and J. Davenport, III. Laramie: University of Wyoming, Department of Social Work.

Wieland, J. S.; Leistritz, F. L.; and Murdock, S. H. July 1979. "Characteristics and Residential Patterns of Energy-Related Work Forces in the Northern Great Plains." *Western Journal of Agricultural Economics* 4:57–68.

Wilkinson, K. P.; Reynolds, R., Jr.; and Ostresh, L. M. 1982. "Local Social Disruption and Western Energy Development: A Critical Review." *Pacific Sociological Review* 25(3):275–96.

Wilkinson, K. P.; Reynolds, R.,Jr.; Thompson, J. G.; and Ostresh, L. M. 1983. "Divorce and Recent Net Migration into the Old West." *Journal of Marriage and the Family* 45:437–45.

———. 1984. "Violent Crime in the Western Energy Development Region." *Sociological Perspectives* 27(2):241–56.

Part III

COMPARATIVE LESSONS AND FUTURE PROSPECTS

9

Synthetic Fuels Abroad: Energy Development in High Energy Dependency Areas

JOSEPH R. RUDOLPH, JR.

Few national policy proposals born in the 1970s with the intent of reducing western dependency on imported energy have been as disappointing in outcome as the synthetic fuels option. Seldom, if ever, has so much capital been committed to a particular objective by so many public and private actors on so many continents in so short a time, only to be so rapidly cut back or withdrawn on such equally short notice with so seemingly little to show for the effort. In Britain, Japan, West Germany, and other energy-importing countries, just as in the United States, the private sector was encouraged, often by the availability of very substantial public assistance, to commit its own capital and expertise to the realization of a common national goal: the development of oil and gas from coal, shale, and other fossil fuel sources. Yet few of these projects are still being pursued, and fewer still have been completed.

This chapter explores the recent rise and decline of the quest for synthetic fuels in the countries of the developed world. Attention is given both to those projects launched by private and public actors inside individual countries (albeit often with outside partners) and to those projects undertaken within the framework of international organizations or under the sponsorship of such bodies. The emphasis is on those projects pursued by countries with far fewer indigenous sources of energy and a far higher level of energy de-

Much of the research upon which this chapter was based was supported by grants from the University of Tulsa and the Faculty Research Committee of Towson State University. The author is appreciative of this support and thankful to those officials of the European Communities, the International Energy Agency, the United Kingdom DOE and National Coal Board, and elsewhere whose availability and assistance made the field work profitable.

pendency than the United States, but who have been no more successful in developing synthetic alternatives to imported oil and gas.

THE PURSUIT OF SYNFUELS IN WESTERN COUNTRIES

The oil crises of 1973 and 1979 placed energy high on the political agendas of most states in the western developed world to at least as great an extent as occurred in the United States and for understandable reasons. With its own four-fuel energy system, the United States was never under the same degree of duress to develop alternative energy sources as other oil-importing states; nor was its level of dependency anything like that of others. At the time of the 1973 oil embargo, oil imports accounted for perhaps 15 percent of America's total energy consumption; by the time of the 1979 oil crisis, oil imports had peaked at slightly over 25 percent of the energy used in the United States and half of the oil consumed in America. At best, or worst, imported oil was important to the good life in America, but was hardly crucial to economic survival itself. For the rest of the oil-importing world, however, oil imports accounted for 45 to 80 percent of total energy consumption, a level of dependency that reflected the pervasiveness of petroleum in housing, industry, commerce, and transportation and the general indispensability of imported oil to the survival of the economic systems of most western states by the 1970s. Consequently, with the exception of Britain and Norway, which were already aware of North Sea oil by 1973 and producing it by 1979, the oil crises provoked a series of crisis decisions. Out of these, in turn, emerged most of those commitments by foreign governments to the development of synthetic fuels. (For an inventory of the synthetic-fuel projects launched throughout the world since 1981 see Doerell, 1982.)

Typical of these commitments were West Germany's Program for Non-Nuclear Energy Research, with its stress on the importance of coal liquefaction in Germany's energy future, and Japan's Project Sunshine, which viewed synthetic fuels as playing a major role in Japan's evolving energy system (*Technologie Programm*, 1979; Sixth Report, 1980; *New and Renewable Sources*, 1981; Schilling and Peters, 1981). Other states, such as Australia, New Zealand, and Canada, which together with the United States, Britain, Japan, and West Germany accounted for 90 percent of all synthetic fuels projects during the last decade, embarked upon programs to encourage foreign technology holders to develop commercial synthetic-fuel plants in their countries (Australia Information Services, 1981; *NZSFCL Background*, n.d.). Indeed, collaboration between public and private actors was a central characteristic of most of the major projects that emerged after the 1979 oil crisis, when the nearly $36 per barrel market price of OPEC oil encouraged a variety of private energy firms to consider synfuels as a potentially competitive energy source. Nevertheless, few of the countries in today's developed world are any nearer today to developing commercial-scale synfuel plants than the United States. What happened?

Obstacles to the Development of Synfuels

Obviously, national differences cannot be discounted in explaining the fate of synthetic-fuel projects throughout the developed world. The most important differences have involved culture and history; prior experience with governmental ownership of energy sources; the nature of the political actors; institutional arrangements including parliamentary government, majority rule, and disciplined party systems; and energy profiles. In this regard, differences in institutional settings have made the decision-making process concerning synfuels much more executive-centric abroad than in the United States. Similarly, the history of public ownership of energy industries in many European states gave the debate over government participation in synfuels a different character outside the United States. Direct aid and participation in synfuel ventures could more easily be discussed as first options, not as alternatives of a last resort. On balance, however, these indigenous factors had far less influence on the fate of synthetic-fuel projects throughout the oil-importing world than those factors that were largely responsible for influencing American efforts to advance synthetic fuels technologies. These include, to some degree in ascending order of importance, the following: (1) the pluralistic pattern of decision making in democratic western systems; (2) the financial problems of funding high-cost projects in recessionary times; (3) the still underdeveloped state of synthetic fuels technologies; (4) partisan political change in oil-importing states during the past decade; (5) the changing nature of the world oil market in the 1980s; and (6) the emphasis on commercialization in R&D designs after the second oil crisis in 1979.

Synfuels and Pluralistic Democracy. The development of public policy in the energy field has been generally difficult for democracies to manage during the past decade. The breadth of interests represented in energy policy making, the general sensitivity of the topic, and the unpleasant, zero-sum nature of many of the available options have, throughout the pluralistic systems of the western world, made the hard choices involved in creating national energy policy very difficult to make. And what has been true of energy policy making in general has been true of energy policy in the area of synthetic fuels in particular. Most western countries have found it difficult to commit themselves to synthetic-fuel programs and to maintain that commitment in the context of changing circumstances.

Energy is an issue that affects far too many, often competitive groups in any society for governments to be able to commit themselves easily to a tidy energy policy or thoroughly to a specific energy option. Too many groups must be somehow appeased for policy making to move rapidly in the energy field in any democratic system. Sometimes the groups are especially well placed inside government, as in Britain, where each energy source has had a public-sector actor representing it during the past decade and where public policy has tended to reflect the changing balance of power between these competitors for political favor and resources. Yet even where vested interests have not penetrated inside the policy-making process, energy policy making must inevitably take into account well-mobilized groups outside of government committed to programs that might have to be scaled back in order to finance costly new ventures. The influence of these groups too will shift over time even in disciplined party-parliamentary government systems, depending on the size of the government's majority and changes in the political environment. Except perhaps in the field of defense spending, pluralism begets conservatism and incrementalism in government, making it especially difficult for polities to undertake and sustain bold new policies except in moments of crisis, when the urge to act overcomes the bias toward incrementalism. Even then, however, the resultant policy is vulnerable to counterattack once the sense of crisis ebbs (Hamlett, 1987; Willis, 1987).

The Problem of Public Financing. Public financing has been another source of difficulty for synthetic-fuel programs throughout the developed world. Synthetic-fuel R&D projects are both full of uncertainties, hence the unwillingness of the private sector to pursue them without some public financing or contributions, and costly, hence the unwillingness of most states readily to appropriate funds for a wide range of projects. In the United States, this problem emerged later rather than earlier in the framework of the deficits currently facing budget authors in the American government. Outside the United States, financial considerations have never favored the development of a government-sponsored industry, especially given the tendency of synfuel plants to race inflationary rates upwards in terms of their cost of construction.

It is one of the odd features of the economics of synfuels that events during the past decade, in particular soaring oil prices, both encouraged governments to develop synthetic fuels programs by making a domestic synfuels industry appear to be potentially competitive with the price of unsure supplies of petroleum from abroad and deprived governments of the funds necessary to commit themselves without reservation to synfuels. Invariably, the high oil prices of 1979 led to stagflation, declining rates of economic growth, higher unemployment as a drain on state resources, and decreased tax revenues as a result of the recession. Indeed, recessions themselves are notoriously bad moments for any state to launch a new and costly program. Hence, in the early 1980s the United States was very much in a class by itself in terms of ability to finance a large-scale synfuels program. It alone had a source of revenue not generally burdensome to taxpayers that could be used to finance massive research and development programs in the energy field: the windfall profits tax.

No other country was as fortunate, except perhaps Britain. By the early 1980s Britain had exportable amounts of North Sea oil to generate revenue for its treasury. On the other hand, the global recessions since 1973 have been especially hard on Britain, regardless of which party has been in office, because much of the country's income has had to be diverted from the North Sea to programs involving social welfare maintenance and economic adjustment. Few funds have been available to develop new energy resources to prepare Britain for the day when North Sea reserves and revenue will begin to run out.

Elsewhere in the West, governments have had no escape from the harsh economic realities of the past decade; it is, therefore, not surprising that in recent years, confronted by recessions at home and declining oil prices abroad, western governments have substantially reduced their previous commitments to the development of synthetic fuels. The United Kingdom's and West Germany's moves are not atypical: Thatcher cutting the National Coal Board's central liquefaction project to a tenth of its original design, and Kohl reducing both the number of pioneer synfuel plants that his government would be willing to support, if any, and the level of support it would be willing to extend.

Technological Hiccups. Closely related to the above, a third factor hindering the development of synfuels involved the synfuel technologies themselves. Although the major techniques of gasification, liquefaction, extracting oil from shale, and processing tar sands were already "known" by the 1970s, few had been tested on a large scale. The only real exception was the indirect liquefaction technique, developed in 1925 and utilized by Germany to make most of its aviation fuel during World War II and later employed by South Africa in building its first liquefaction plant in 1950. Scaling these technologies up from laboratory bench-scale operations that processed only very small amounts of feedstock per day to larger demonstration facilities raised a variety of technological problems ("hiccups" in the industry's reports), socioenvironmental problems, and financial difficulties. Thus, cost overruns had become another central feature of synfuel R&D by 1980 and the justification most frequently offered by industries that canceled their synthetic fuels projects by 1982 (Lefevre, 1984).

Political Change. A fourth factor, more coincidental than the others, but equally destructive to efforts to develop a synfuels industry in key western countries, involved partisan political change. The effect of the election of Ronald Reagan on the development of the American synfuels corporation is examined elsewhere in this volume (Willis, 1987). In a similar manner, the strong support of the National Coal Board's liquefaction project in Britain was undone in 1979 when the Labour party was replaced in office by a government with a decidedly free-market philosophy about energy development as well as economic policy. Meanwhile, the commitment of the Social Democrats in West Germany to an ambitious synthetic fuels plan was tabled when the Kohl government took office because it conflicted with both Kohl's market philosophy and his austerity program for the German economy. The fall of the Liberal party had serious implications for Canada's synfuel plans, and in New Zealand the opposition party found the government's support of Mobil's M-process plant vulnerable to political attack in the oil-glut era of the eighties. By themselves, these changes might not have been lethal, but in the context of the changing international energy market, partisan political change removed or exposed the vulnerability of synfuel advocates at the very moment when these projects most needed strong public backing.

The Changing Global Energy Market. Interest in synthetic fuels has ebbed and flowed historically in other countries much as in the United States (Vietor, 1980). Tight energy markets and rising oil costs have historically stimulated R&D in the synfuels area; gluts of oil and the falling oil prices that have usually followed these shortages have usually ended the quest for synfuels by making them appear economically unprofitable. So it has been during the last decade. The first energy crisis stimulated interest once again in synfuels in the western oil-importing world, albeit largely at the governmental level. The 1973 oil crisis was not only a crisis involving a fourfold increase in the price of OPEC oil, which made synfuels and other alternative energy sources appear more attractive, even if they remained more expensive in the short term than the new price of OPEC crude, but also a crisis involving the security of supply of the energy source upon which western economics and military establishments ran. Consequently, national security considerations dominated the early discussions of the desirability of coal liquefaction, shale oil, and tar sands petroleum as alternatives to the liquid energy source supplied by the oil-exporting world. Still, between 1973 and 1979, the costliness of major synfuels projects combined with the unwieldy nature of decision making in the oil-importing world, the continuing uncompetitive cost of synfuels, and the apparent stabilization of OPEC's oil prices by 1974 to prevent even those states that were spurred by national security considerations from going beyond discussing the synfuels option to diligently pursuing it.

It was thus the 1979 oil crisis that provided the big boost to public support for and private involvement in large-scale synthetic fuels projects. To governments, the sudden redoubling of the price of oil to nearly $36 a barrel as a result of the fall of the shah of Iran and the subsequent disruption of Iranian production made the development of alternatives to OPEC oil seem even more imperative. Otherwise, it was widely believed, the best-laid plans for economic management at home could be undone by frequent, sharp increases in the cost of the imported petroleum upon which western economies depend. Thus, a subtle but important shift from national security to an economic justification for synfuel ventures took place, which was to be important later. To the energy industry, the increase in the price of international petroleum seemed to bring closer the day when synfuels would be able to compete against natural crude. Hence the decision by governments to increase their funding of liquefaction and other synfuel sources, and the decision by industry to mount ambitious R&D projects. Within a year, the twelve largest international petroleum corporations were committing more than $1 billion to a variety of synfuel projects, with more to come, and, in many instances, without governmental assistance or participation in the projects.

The sudden, nearly $20 a barrel addition to the cost of imported oil in 1979 created a global recession that in turn dampened western demand for oil at the same time that the high cost of oil was stimulating the exploration for and development of oil outside of OPEC around the world. By the early eighties an excess of oil was already appearing on the world's market and the price of OPEC oil was falling. In this context the emerging cost overruns in the synfuels projects, already under way, assumed major importance. Had the price of oil continued to go up—it was once predicted at $50 a barrel for 1985—the cost overruns might not have been fatal. But with the price of oil dipping below $30 a barrel and the projected cost of synfuels from first-generation plants being pegged upwards of $95 a barrel for some synthetic-fuel sources, the bailout began. By summer 1982 Exxon had cancelled its $5 billion shale-oil project; a consortium of several Canadian oil companies had withdrawn from Canada's $11 billion Alsands syn-oil project, despite a last-minute offer by the government of major assistance to the project; and British Petroleum had backed out of Britain's more modest, $55 million Point of Ayr project, leaving it in limbo.

The R&D Design of Synfuel Projects. If the synfuel projects in the west had remained rooted in national security arguments, the downturn in the price of imported oil in the early 1980s would probably not have been nearly as fatal to synthetic-fuel R&D as it was. Governments would probably have favored direct funding of projects, providing them with some insulation from changing market conditions. However, given the costliness of these projects and the growing concern with fuel costs in the post–1979 period, governments welcomed, from the outset, the participation of private-sector partners in synfuel demonstration projects. In doing so, these same governments at least tacitly accepted the suitability of the fuel source for commercial development as the principal criterion for evaluating a project, rather than its impact on the environment or its value in reducing a state's level of energy dependency.

The efficacy of using private corporations as vehicles for the realization of national goals in the development and demonstration of new technologies, without at least a secondary backup system, is open to question. The projects most likely to be pursued will be those most financially attractive, not necessarily those best utilizing a country's resource base. Oil-shale projects, thus, have been more quickly pursued than coal liquefaction projects, because they have been consistently viewed as the projects promising to be the least expensive per barrel of synthetic fuel, although coal is a much more generally available feedstock for countries seeking to reduce energy dependency through the production of synthetic substitutes for OPEC oil ("Economic Comparisons," 1981). Similarly, gasification projects have been a more likely bet for commercialization than liquefaction and tar sand technologies. Hence, gasification projects are still being pursued today, despite the fact that they provide no alternative to OPEC for liquid fuels, while virtually all other synthetic-fuel projects beyond the pilot-plant stage have been canceled everywhere in the oil-importing western world except South Africa. But most importantly, by allowing all projects to turn on private-public partnerships rather than opting for the GOCO (Government-Owned, Corporate-Operated) or GOGO (Government-Owned, Government-Operated) approach to some R&D projects, the West allowed all major synthetic-fuel demonstration projects to fall hostage to changing market circumstances and to cancellation when the private parties in them chose to withdraw from the ventures.

The effect of this factor and the others on synthetic-fuel projects can be read in the history of these ventures in virtually every western country. The case study of Britain's pursuit of coal liquefaction offered here is typical.

The Point of Ayr Synthetic Fuels Project: A Case Study

While the British Gas Corporation has carried the ball in Britain since 1973 in the area of coal gasification, liquefaction efforts in the United Kingdom have emanated from the National Coal Board (NCB) and its attempts to advance its laboratory work in converting coal to transportation fuel. The tests themselves predated the 1973 energy crisis, but were small-scale operations not unlike the nominal, "keeping the hand in" type of research to be found occurring in many states prior to 1973, sponsored by public and private organizations alike. Thus, the £40 million, five-year research program for the study of direct methods of coal conversion that the NCB proposed following the first oil crisis called for a major change in the board's R&D activities from bench-scale research processing 100 pounds a day of feedstock to the construction of a pilot plant capable of operating at the 25-tons-a-day level. Getting the proposal through the system, however, proved to be difficult, especially since by the mid–1970s Britain could anticipate middle-term energy independence via the North Sea's rapidly developing oil fields.

Even the coal-friendly Labour government elected in 1974 did not formally agree to support the development of a coal liquefaction pilot-plant program in the United Kingdom until five years later in the summer of 1978 (*Coal Technology*, 1978). Thereafter events seemed to move rapidly. By mid-1979 British Petroleum had agreed to join the project, and additional capital was expected from the European Communities, leaving the government responsible for contributing only two-thirds of the total cost it had promised. Shortly thereafter, the Point of Ayr colliery in northern Wales was selected as the site for two pilot plants, then estimated to cost £35 million to construct.

Yet appearances were misleading. These developments represented the continued momentum of decisions made under the Labour government. But in the spring of 1979 Britain elected a Conservative government with a decidedly different view on energy policy in Britain, one far more favorable to market mechanisms and less favorable to government intervention in the development of new fuel sources. Indeed, by 1980 the official view of the government was that, given the then-projected cost of £160 million to construct the two plants and the generally high cost of British coal, the United Kingdom with its North Sea oil was likely to be one of the last industrial states to develop a liquefaction industry (U.K. DOE Officials, 1983). The prediction has not been proven wrong to date. Not until May 1981, three years after the Labour government pledged £20 million to a £35 million program, did the Thatcher government agree to assist the development of a coal liquefaction process in Britain. Its offer was not a generous one. Taking the position that the NCB's "technology can be successful only if it is developed commercially," Under Secretary of State for Energy John Moore told the House of Commons that the government would contribute £5 million to the building of a facility at the Point of Ayr if the NCB could raise elsewhere the remainder of the £50 million then estimated necessary to construct a dual-process, single 25-tons-a-day pilot plant (*Hansard*, May 22, 1982).

Thereafter, the chances for the project being completed deteriorated rapidly. The NCB subsequently agreed to limit its plans to a smaller plant testing only one process. But the following summer even that £35 million plant slipped out of reach when both of the board's prospective private partners, Phillips Petroleum and British Petroleum, withdrew from the venture, citing the need to reduce their R&D expenditures in light of the oil glut and their declining profits. By the fall of 1982, the project was officially abandoned on the ground that private-sector support for it had collapsed. The pilot plant had become a casualty of partisan change, market forces, and an emphasis on commercial development by the government and private-sector partners alike.

The NCB still pursues its liquefaction project, but since 1982 it has been further downscaled to a 2.5-ton-a-day, semi-technical plant generally deemed too small to answer the questions necessary at a pilot plant before building larger demonstration plants to test the technology at near-commercial level. Furthermore, even at this level it has had trouble in meeting the government's requirement that it attract a private partner to demonstrate commercial interest in the project and EC backing to get a 10 percent contribution from the government. The date for ground breaking has been steadily advanced as estimates of construction costs, now at £35 million for the 2.5-ton-a-day facility, have risen, and it was only in March 1986 that the project was officially "launched" by the government.

Much the same tale prevails elsewhere. Australia's National Energy Research, Development and Demonstration Program assumed from its origins a decade ago that private enterprise would be necessary for the projects to succeed commercially, but private partners have been hard to attract since the price of oil began to decline. Similarly, since launching its Project Sunshine in 1974 to reduce its dependency on foreign oil from 75 percent to 50 percent within ten years, Japan has had trouble retaining commitments to the synfuels projects it has considered. The fates of synfuels projects in the United States, Canada, West Germany, and New Zealand have already been mentioned. Thus, only in one country in the oil-importing world, the Republic of South Africa, has it been possible to say today that synfuels have become an important part of the state's energy profile since the oil crises of the seventies.

The South Africa Exception

South Africa's experience with synthetic fuels dates from 1947, when a public decision was made to produce oil from coal. With the state providing the capital, SASOL (the South African Coal, Oil, and Gas Corporation) was registered in 1950 and began selling gasoline five years later from its SASOL I plant with a production capacity of 5,000 barrels a day minimum. Still, there as elsewhere, it was the oil crisis of 1973 that provided a big boost to the country's synfuel industry. In order to reduce the country's vulnerability to the type of political blackmail the West experienced under the Arab boycott of 1973–1974, the government committed itself to two additional SASOL plants, each ten times larger than the first. The second oil crisis encouraged the government to accelerate the timetable for their completion. Along the way, SASOL was restructured in 1979 and went public in 1980. The financing of the synfuels experience relies on a state subsidy, approximately a tenth of the gasoline's $2 a gallon price in the 1970s, and a substantial amount of the corporation's financing is still public, derived from South Africa's gasoline tax.

All three plants utilize the Fischer-Tropsch process of indirect liquefaction, and their combined output provides South Africa with a substantial share of its oil needs. How large a share is guesswork. Figures pertaining to the precise production of these plants and their contribution to the country's liquid fuel needs are not available for reasons of "national security." Yet it is precisely South Africa's attention to political-military considerations in building and expanding its synfuel production that explains the success of its synfuels venture. The country's cheap coal, inexpensive labor, and lax environmental code are not unimportant. But its willingness to view the plants almost exclusively from a strategic standpoint and to support them with subsidies when necessary, rather than to apply commercial-economic tests to their operation, has made South Africa's experience with synthetic fuels during the past decade unique among oil-importing states.

This is not to say that South Africa's synfuels industry is invulnerable to criticism. Given the "bargain" price of oil on the world market today, paying the price of reduced dependency is costly, and in almost any other political environment one might expect to hear the type of arguments currently being voiced in Brazil against its government's decision to switch to high-cost, indigenously produced alcohol fuel. However, the political structure and governmental agenda in South Africa do not permit discussion of very much in purely economic terms. Thus, South Africa's experience with synfuels in this area illustrates that synfuel technologies can work, but so far have only been developed on a large scale in circumstances difficult to duplicate elsewhere in the developed world.

MULTINATIONAL COOPERATION AND THE PURSUIT OF SYNFUELS

Given the high risks and costs involved in developing and demonstrating the preliminary, commercial-scale plants employing synthetic fuels technologies, cost-sharing ventures make considerable sense. Substantially for this reason, the dominant format for synfuel R&D inside the countries of the western world has involved collaboration between private corporations and the public sector.

The same cost-sharing logic should also encourage other forms of collaboration: corporate partnerships in the absence of public funding, binational and multinational research consortia, and even private and public partnerships under the auspices of international organizations. In examining the traditional assumptions behind international collaboration in the developing new technologies, Nau and Lester have concluded that the costs involved and the scale required to utilize and market these technologies require international cooperation and cooperative frameworks (1985). Other conventional arguments in favor of such cooperative arrangements revolve around the avoidance of duplicative work and the establishment of projects best reflecting the comparative advantages of host countries (Bobrow and Kudrle, 1979). Nevertheless, despite the weight of these arguments and the efforts of such states as Japan and such international bodies as the European Communities, the efforts to achieve multinational, multicorporate, and international cooperation in the development of alternative energy resources have never been very successful.

Obstacles to International Collaboration

Cooperation among individual countries in the energy field in general has been persistently blocked by obstacles ranging from the nature of the international decision-making arrangements to nationalism and legalism. During much of the postwar period, because of the western emphasis on economic growth and the frequent existence of a world market with a glut of cheap oil, little attention was given to energy by oil-importing states. This tendency is best reflected in the casual interest paid to OPEC as it evolved in the 1960s or even later when the oil market began to tighten at the end of the decade and the western oil companies began to lose their control over it. Consequently, at the time of the oil crisis, there was no precedent or international institution available to assist the West in integrated energy policy making. As Willis has observed, international energy policy is in many ways as much a post–1973 phenomenon as is national energy policy making in the West (1985).

In the absence of such international machinery, western countries established during the darkest hours of the 1973 crisis the broad nationalistic framework of subsequent western responses to energy matters when they rushed into the spot market to engage in panic bidding against one another for available oil. Although perhaps lamentable, it was an understandable reaction. Since these international energy crises manifest themselves in national problems, even national security problems, they must be treated in a national framework, even though national assessments of national needs, perhaps mixed with a little national pride in national technologies, have saved the day. Such matters are too important to individual states to be dealt with through arrangements that depend on the cooperation of others.

The genesis of this mindset can be found well before the 1973 oil crisis affected the postwar western world. Decades ago France and Italy created their own national petroleum companies and encouraged them to compete with the major multinationals rather than rely on the established Seven Sisters cartel to supply French and Italian petroleum needs. Similarly, most studies suggest that in the field of nuclear energy the European states, as soon as they had sufficient resources, side-stepped cooperation within EURATOM in favor of the development of individual national nuclear energy, even at the cost of duplicative R&D. What has been generally true of the development of energy in the international system has also been true in terms of international cooperation on synthetic fuels. The paradox persists: energy security is too important a subject for states to risk working at cross-purposes with one another, but national security is too important to individual states for them to permit others to define it for them (Rudolph and Willis, 1985).

International cooperation in the area of synthetic fuels R&D has thus been possible only on small-scale projects. The larger ones involving demonstration plants, by contrast, have tended to be concerned with demonstrating a specific synthetic-fuel technology for the benefit of a single country. Consequently, they have on occasion tended to be redundant when different countries have committed themselves to testing processes that have involved minor variations on the same technology. The one major exception to this rule, the large-scale, SCR-II liquefaction project linking the United States, Japan, and West Germany, collapsed as a result of a national action: Reagan's decision to cancel the DOE's planned participation in it. Thereafter, Japan and West Germany did not have any interest in trying to revive the project by appealing to the Synthetic Fuels Corporation for the missing U.S. financial contribution to a U.S.-based project designed to demonstrate the technology of a major American petroleum company; nor did other governments, in general, show much interest in cooperative ventures, especially with the United States.

Finally, nationalism in the guise of legalism frequently obstructed international cooperation in the synfuels field, again with the United States leading the way. American Synthetic Fuels Corporation officials, when they attempted to reassemble those synfuel projects shattered by the withdrawal of their principal corporate financiers by soliciting assistance for those projects from foreign governments, met the immediate opposition of congressional spokesmen who argued that the SFC was chartered to advance the development of an American synthetic fuels industry and, under the Energy Security Act, could not permit foreign governments and industries to participate in SFC-funded projects, except perhaps as minority partners ("Bids for Foreign Funds," 1982).

International Institutions and International Collaboration

International organizations have enjoyed no more success than individual countries in launching synthetic fuels projects or in encouraging national actors to collaborate in the demonstration of synthetic fuels technologies. Some of these bodies have simply been ill suited to deal with the topic. NATO, for example, is not constituted to deal with energy matters, despite the national security implications of energy policy in the Atlantic Alliance and the weakening of NATO unity by the recent fights among its members over energy issues (Miller, 1983). Yet even those agencies explicitly created to function in energy areas, such as the European Communities and the International Energy Agency, have not been especially successful in facilitating cooperative international ventures in energy R&D.

The European Communities. Like NATO, the European Communities has had trouble coping with the new energy issues of the post–1973 world. The oil crises and their repercussions have aggravated existing rich/poor–big/little strains within the EC and have added a new cleavage between the Communities' energy haves, basically those members with access to North Sea oil and gas, and the have-nots. The crises also transformed energy into an issue much more likely to be treated by the Communities' Council of Ministers, where the views and interests of member states are guarded, than by the Eurocrats staffing the EC's commission. As a result, matters like the development of new energy sources have often been stalled by the split on the council between those ministers representing governments with a free-market approach to energy development and those with an interventionist philosophy.

More specifically, the fund created by the EC in 1978 to provide "financial support for demonstration projects to exploit energy alternatives" is controlled by the council, and funding awards frequently go to projects as much on the basis of their political merit as their technological worthiness for support (EC Officials, 1983). The commission controls only a very small portion of the EC's R&D budget and in administering it must adhere to the

EC's general guidelines, including the requirement that funded projects have "industrial potential" and "commercial viability." Given the present level of oil prices and synthetic fuels technologies, these requirements would exclude EC participation in most synfuel projects, if narrowly interpreted.

In practice, the legal obstacles to cooperative R&D ventures can be overcome through trade-offs and by broad interpretation by the commision of its charge to aid in the development of technologies beneficial to member countries. Unfortunately for the cause of energy policy making, the EC cannot overcome the fact that it does not actually function as a supranational body in the development of alternative energy sources. The EC can react to the requests of member states for assistance, but it cannot inaugurate synfuel or other technology-developing projects. The EC can subsidize demonstration projects within member states, but it cannot assume control over these projects or continue them when they collapse at home for want of support by the national government or private corporations.

The International Energy Agency. As a vehicle for developing alternative energy sources, the IEA is even more limited than the EC. Although created explicitly to respond to the problem of energy supply in western countries in the aftermath of the first oil crisis, the agency functions as a confederate addendum to the OECD with a voting system that tends to produce a diluted, common-denominator approach to decision making (Bobrow and Kudrle, 1979). Even as an alliance its effectiveness has been undercut from the outset by the nonparticipation of France in the operations of this Paris-based body.

At its best, the IEA can serve as an international library for energy information, a forum for debating energy options and policies in the oil-importing world, and a source of advice to its members on policy-making matters. As a source of counsel, it has frequently urged its members to develop synthetic-fuel systems as a hedge against a return to the oil-scarce energy crisis days of the seventies (IEA, 1982). However, lacking both political autonomy and a research budget of its own, the IEA is unable to encourage such programs through direct contributions. Moreover, the one service it does provide, a treaty format for the pursuit of multinational projects, has been largely rejected by western countries as too rigid an approach to developing unproven technologies. The latter argument is probably valid, but it is more interesting as a reflection of the major obstacles that the IEA and other international organizations face in trying to create multinational ventures for the development of new energy sources, including the already-discussed tendency of national actors to see energy problems in national terms, the desire of individual countries to develop and perhaps hoard successful national technologies, the inclination of national leaders to subject the policies of others to their own tests, and most important, the absence of any shared belief that multinational ventures can be more successful than national responses to national energy problems (Miller, 1983, p. 480).

In short, although the IEA has achieved some success in initiating multinational programs in such areas as coal utilization, oil stockpiling, and oil sharing, it remains essentially an assembly for the discussion of policies rather than a facilitator of collaborative ventures. Even as a forum its effectiveness has been undercut with the passage of time. As memories of the events of 1973 and 1979 have dimmed, the IEA has increasingly found it difficult to persuade member states to consider long-term energy needs and to develop alternative energy sources, especially in partnership with other members. Thus, collaboration among IEA members continues to occur in the energy field on an informal basis, with few national actors involved in any project, and with the projects limited to preliminary bench and pilot-plant stages of research.

Private Corporations and International Collaboration

What is true of the national actors has been true of corporate actors involved in synfuel research during the past decade. Information sharing, collaborative ventures, and cost-sharing arrangements exist. Not only were all of the Seven Sisters and many of their international petroleum cousins actively involved in joint synfuel projects during the last decade, but so were a large number of public-sector industries. Ruhrkohle, for example, was a participant in an Australian coal-mining project and liquefaction projects with Exxon and in America with Ashland Oil as well as in West Germany's major liquefaction project. Nevertheless, the rule seems to be much the same as that governing the behavior of states. The larger the project and the nearer to commercial use, the less collaborative it is likely to be. Similarly, projects sponsored by international organizations appear to be viewed by the private sector as the least desirable of all because of the danger that in such a format the information generated today will be widely disseminated to potential competitors of tomorrow.

SYNFUELS IN A COMPARATIVE CONTEXT: CONCLUDING OBSERVATIONS

If old soldiers are always fighting the last war, old (and young) political scientists are forever examining the last crisis and trying to wring from it as many lessons as possible. The dangers of this way of thinking are obvious: it can produce more lessons than any crisis can support, or it can generate the wrong lessons. It is nonetheless difficult to review the comparative and international efforts of western countries to advance synthetic fuels technologies during the last decade without concluding that there must be better ways to pursue technology, make public policy, and administer government programs in this field than those that were chosen. This is not to say that all of the projects were for naught. Some still continue on a large scale, essentially gasification projects, and some of the data derived from laboratory experiments may yet prove useful in the future, for example, the NCB's work in boosting the octane of synfuels for use in aviation (Davies, 1983). However, despite numerous projects and billions of dollars of research and development expenditures, most of the major technological, economic, environmental, and social questions contained in the synfuel option remain unanswered. We still do not have demonstration-size plants on line to illuminate the costs and advantages of commercial-scale facilities for producing oil via direct liquefaction techniques, tar sand extraction, and oil-shale processing.

Nor do the patterns discernible in the western world's most recent flirtation with synthetic fuels encourage one to believe that future events will lead to different outcomes. No country wants to leave a matter as important as energy development to others in moments of crisis; therefore, there is little incentive to press for multinational cooperation when energy issues are most salient. On the other hand, when the crises subside, national programs tend to be canceled. Similarly, the private sector is equally unwilling to pursue multicorporate and multinational projects in the area. When the technology appears competitive, no technology holder wants to share the potential profits with others. But when synthetic fuels no longer appear competitive, no one wants to continue large-scale R&D projects in the synthetic fuels area.

Indeed, the only enthusiasts for cooperation in synfuels development are perhaps those countries with little prior work in the area, but substantial deposits of coal. These countries, who frequently court the states and corporations with the technological experience, are also in the forefront in proposing collaborative ventures. But even if launched, these projects, too, tend to be canceled when the air of emergency passes. Finally, nowhere does the indirect-subsidy-by-government approach seem sufficient to preserve projects slated for termination because of changing economic circumstances. Only in South Africa has the government been willing to challenge the unknowns and commit itself to develop a synthetic fuels industry.

In the context of these troubling patterns and the absence of any major progress in answering the questions associated with the synthetic fuels option during the past decade, I offer these concluding reflections with respect to synthetic fuels and policy making in the contemporary oil-importing world.

Synthetic Fuels and Future Energy Mixes

Synthetic fuels are not going to play any role in the immediate future in the energy system of the United States or any other country except South Africa. Political and economic factors have stalled the development of synthetic fuels virtually everywhere at approximately the place they were a decade ago. There may even be some lost ground. The failure of political processes to realize the goals to which they explicitly committed themselves has frequently been interpreted as evidence of the infeasibility or impracticality of synthetic fuels under any circumstances. These conclusions are as premature as were the optimistic forecasts for synfuels of only a few years ago. They do, however, make public financing of large-scale synfuel demonstration plants more difficult and unlikely, as do the current disarray inside OPEC, the contemporary complacency about energy as a result of the recent collapse of OPEC oil prices, and the availability of large amounts of oil on the world market from non-OPEC sources. But if western democracies were to reconsider developing synfuel technologies, their recent experiences in the field suggest some of the directions that public policy makers might want to consider.

First, the justification for public-sector support of synfuel R&D must be strategic; that is, it must be based on a desire to enhance national security by developing additional domestic supplies of liquid fuel. Whether synthetic-fuel programs could be that helpful to states with extremely high levels of dependency on foreign oil for their energy is an open question, but most countries do not have Japan's level of dependency (nearly 80 percent in 1973), or America's staggering appetite for millions of barrels of imported oil per day. For them, a modest number of functioning synfuel plants would have a more than trivial effect on energy supply. Perhaps South Africa provides a singular model for western states to follow: build synthetic-fuel facilities for political reasons. If they prove to be commercially successful, then by all means sell shares in them or sell them off to be operated by the private sector, but build them.

It is unlikely, except in the most drastic of future energy scenarios, that economic conditions alone will encourage the public sector or private enterprises to invest the very substantial capital necessary to construct pioneer commercial-scale plants. In the near future, imported oil is apt to be much less expensive than tomorrow's synfuels. Indeed, in predicting future price comparisons between synfuels and OPEC crude, forecasters may prefer caution on the side of oil, if only to avoid being caught a second time in the error of overestimating the future price of petroleum. On the other hand, western countries can always use a more secure source of fuel to meet their military needs. If synfuels can be developed and sufficiently subsidized to be offered to the military at a price near that of imported oil, they are not likely to want for purchasers.

Second, the program should be a small-scale one that provides funding to a limited number of projects. The goal should not be the development of a synthetic fuels industry but demonstration plants to test several technologies and, in doing so, to produce 100,000 or more barrels of oil per day for military use. In short, the program should lie somewhere between symbolic action and a politically visible and potentially vulnerable, multibillion-dollar program available to virtually every small energy entrepreneur in a country. Smaller and better-defined programs should also fare better in the pluralistic setting of democratic policy making. Research and development projects defined to demonstrate synthetic fuels technologies could even enjoy a slight advantage over policies promoting other energy alternatives. Unlike solar energy and other renewables, synthetic fuels do not challenge existing fuel systems, but complement them. A developed synfuels industry can stimulate coal industries and bring profits to the gas and oil industries, which are currently in the lead in synfuels research and usually the principal private-sector owners of shale and tar sands reserves.

Third, a synfuels demonstration program should be based from the outset on a firm, upfront economic and political commitment to see the program through to the pioneer-plant stage in order to learn the technology's costs and benefits, even if doing so means that the government must accept 100 percent of the costs. A wide range of private-public collaborative arrangements should be studied, but if the goal is to guarantee the construction of synfuel demonstration plants, the GOCO formula may have to be accepted as the primary means toward that end. Inducements such as public subsidies and loan guarantees provide no public guarantees or project guarantees. They lead to reliance on the private sector for project completion, which in turn injects commercial considerations into an R&D program and endangers it whenever economic conditions encourage the private industries to withdraw from the project. The NCB's and the SFC's inability to persuade private corporations to invest in demonstration plants in an era of falling oil prices attests to the difficulty of launching or sustaining demonstration programs under adverse economic conditions with only indirect public assistance of the loan guarantee or price support variety. It is no coincidence that the only country since World War II to develop a commercial-scale coal liquefaction industry is the one country to have undertaken the project almost exclusively on the basis of national interest considerations and to have built an industry entirely with public financing.

Administration, Cooperation, and the Development of Energy Technologies

Whatever the financial arrangements, the construction of pioneer synfuel plants will probably involve cooperation between the public sector with its financial resources and the private sector with the technological expertise necessary to demonstrate synfuels technologies. Cooperation with other countries or the industries of other states would also provide useful ways of cost sharing. Finding the proper administrative arrangements for these partnerships will not be easy because the western world's most recent experience in the synfuels field suggests that the administration is best that is least intrusive in the research portions of the project and least visible in the political process.

International cooperation based on bilateral and multilateral arrangements involving relatively few national and corporate participants is thus to be preferred over arrangements undertaken within the framework of an international organization. The former affords more protection for the confidentiality of information than the latter; it also provides greater flexibility, including the option of easy withdrawal from joint ventures, a continuing precondition of international collaboration. Ad hoc bilateral and multilateral agreements also require the accommodation of fewer interests in research and development programs and, if the project succeeds, have the "advantage" of limiting access to the new information to only the countries and corporations directly participating in the project. In short, the model project would look very much like the SCR-II venture with an ad hoc administrative system guiding it.

Domestic arrangements would necessarily have to be more formal because democracy requires that the expenditures of public funds be monitored and execution of projects overseen. Nevertheless, there is no need for special agencies to attract attention to themselves. Small, selective research, development, and demonstration programs involving synfuels technologies should be manageable on the basis of contractual arrangements negotiated through existing departments of energy or defense.

In the final analysis, the administrative arrangements are far less important than the will to define projects that can be technologically realized, not those that are merely a political response to an impending energy crisis. The mobilization of that kind of will is something the western democracies have yet to achieve in the field of synthetic fuels.

REFERENCES

Australia Information Services, Department of Administrative Services. July 1981. *Energy Politics*. Canberra: Australia Information Services.

"Bids for Foreign Funds for Synthetic Fuels Leaves Program Open to Political Attack." December 2, 1982. *Wall Street Journal*.

Bobrow, Davis, and Kudrle, Robert T. Spring 1979. "Energy R&D: In Tepid Pursuit of Collective Goods." *International Organization* 33:149–75.

Coal Technology: Future Developments in Conversion, Utilization, and Unconventional Mining in the United Kingdom. May 24, 1978. London: Coal Industry Tripartite Group.

Davies, G. O. 1983. "The Preparation and Combustion Characteristics of Coal Derived Transport Fuels." Cheltenham, England: National Coal Board.

Doerell, Peter E., ed. 1982. *World Synfuels Project Report*. San Francisco: Miller Freeman Publications.

"Economic Comparisons of Four Synthetic Fuels." June 1981. *Oil and Gas Journal* 27:122–24.

European Communities Officials. June 1, 1983. Interviews. Brussels, Belgium.

Hamlett, Patrick W. 1987. "Technological Policy Making in Congress: The Creation of the U. S. Synthetic Fuels Corporation." This volume, chapter 3.

Hansard. May 22, 1982, p. 569.

International Energy Agency. 1982. *Coal Liquefaction: A Technology Review*. Paris: Organization for Economic Cooperation and Development.

Lefevre, Stephen R. 1984. "Using Demonstration Projects to Advance Innovations in Energy and Communications." *Public Administration Review* 44:483–90.

Miller, Linda. 1983. "Energy and Alliance Politics: Lessons of a Decade." *The World Today* 39:476–84.

Nau, Henry R., and Lester, James. 1985. "Technological Cooperation and the Nation-State." *Western Political Quarterly* 38:44–69.

New and Renewable Sources of Energy: Contributions by the Federal Republic of Germany. 1981. Bonn: Federal Ministry for Research and Technology, Press and Public Relations Division.

NZSFCL Background: The Gas to Gasoline Project. N.d. New Zealand Synthetic Fuels Corporation, Ltd.

Rudolph, Joseph R., Jr., and Willis, Sabrina. 1985. "The Politics of Technology, Public Policy, and Administration: The Synthetic Fuels Venture in Western Democracies." Unpublished paper.

Schilling, H. O., and Peters, W. September 1981. "Present Status of Technology for Coal Liquefaction and Gasification in the Federal Republic of Germany." Paper prepared for presentation in Tokyo.

Sixth Report of the Federal Government on Research. 1980. Bonn: Federal Ministry for Research and Technology, Press and Publications Division.

Technologie Programm: Energie des Landes Nordrhein-Westfalen, Mai 1979 mit Fortschreibung Jul. 1979. 1979. Minister—für Wirtschaft, Mittelstand und Verkehr des Landes Nordrhein-Westfalen.

United Kingdom Department of Energy. May 24, 1978. Press Notice Ref., no. 168. London: U.K. DOE.

———. May 12, 1983. Interviews with U.K. DOE officials.

Vietor, Richard H. K. Spring 1980. "The Synthetic Liquid Fuels Program: Energy Politics in the Truman Era." *Business History Review* 54:1–34.

Willis, Sabrina. 1985. "Western Attempts at International Energy Policy Making: The European Community and the International Energy Agency." Unpublished paper.

———. 1987. "The Synthetic Fuels Corporation as an Organizational Failure in Policy Mobilization." This volume, chapter 4.

10

Prospects of Synthetic Fuels in the United States: Past Lessons and Future Requirements

THOMAS J. WILBANKS

The past decade has taught us several lessons about the future of synthetic fuels in the United States. First, synthetic fuels will not become a major part of the U.S. energy picture very rapidly because the management challenge is simply too great even if the fuels promise to be competitive in energy markets. Second, synthetic fuels will not come into widespread use just because government says it wants them, because business must see that there is money to be made over a long period of time—and governments have been known to change their minds much more quickly than that. Third, synthetic fuels are not, however, likely to come into widespread use in the United States in the next generation without government encouragement, because most current views of the future indicate that synfuels are bad business risks.

These lessons amount to a constraint compounded by a predicament. The constraint is that any buildup to a sizable commercial synfuels industry will take some time. Clearly, the kinds of time frames envisioned by the Energy Security Act—reaching 500,000 barrels per day of production in seven years and 2,000,000 barrels per day in twelve—were fanciful, even if the national resolve had been deeper than turned out to be the case. For large, expensive facilities based on largely untested technologies, the process of finding sites, borrowing money, building and debugging plants, and resolving a host of controversies will be time-consuming for the pioneer plants that serve as guinea pigs for the private and public sectors alike. And the prospects of a full-fledged industry will depend on information from this first generation of experiments, unless government is prepared to assume nearly all of the risks.

The predicament is that serious development along these lines cannot proceed without staunch support from major private firms, led by multinational energy companies, but they are almost certain not to head in this direction in the next decade or more without a kind of public policy environment that is hard to imagine: stable government support for synthetic fuels in spite of frequent changes in Congress and the White House, pressures to reduce budget deficits, and political risks in working hand-in-hand with the firms involved. Not only do relatively low oil and gas prices make synthetic fuels look too expensive to sell or buy in the foreseeable future, but the whole idea of synthetic fuels has been "tarnished" by the recent experience of stop-and-go government policy (Landsberg, 1986). If the nation decides at some point that it wants to try again, why should industry be anything but skeptical about the national resolve? How can the nation itself sustain its commitment long enough to get results?

In short, it is difficult to be very optimistic about the prospects of synthetic fuels in the United States in the next quarter of a century, in spite of the facts that abundant resources and workable technologies are on hand and the need to find replacements for petroleum and natural gas, sooner or later, is clear. Whether or not this collective caution is desirable can be debated. Reasonable arguments can be offered for and against. But our experience with synthetic fuels is symptomatic of a larger underlying issue: whether our modes of decision making in this country are capable of preparing us technologically, economically, and institutionally for changed conditions in the long term.

As a foundation for considering this larger issue, the following sections of this paper will first summarize the pros and cons of accelerating the development of synthetic fuels in this country. It will then consider a few different scenarios of decision-making conditions in the next 20 to 40 years to see how the prospects of synthetic fuels might be affected, and it will draw some conclusions about possibilities, probabilities, and policy directions worth considering. Finally, it will offer some thoughts about the more general issue stemming from the synfuels experience.

THE PROS AND CONS OF SYNTHETIC FUELS

From a public policy standpoint, the basic question about synthetic fuels is whether laws should be enacted, institutions created, rules defined, incentives established, or other steps taken to accelerate the development and use of these fuels in the United States. Is there some net social good to be realized beyond what will happen by private initiative alone with no public-sector encouragement?

The answer revolves around an evaluation of the pros and cons of synthetic fuels and an assessment of how the benefits and costs from synfuels would be distributed. To readers of this book, the litany is a familiar one by now. Why worry? Why not just let events take their course? Summarizing very briefly, the case for synthetic fuels rests mainly on five arguments.

First, the world is running out of petroleum and natural gas. The question is not whether we will eventually need replacements for petroleum fuels and natural gas. The only guess is when. As these natural liquid and gaseous fossil fuels get increasingly scarce—and they must because they are being consumed faster than they are being replaced—their prices will rise until alternatives are required to prevent economic stress and political instability.

Second, the remaining coal resource of the world in general and the United States in particular is immense. Estimates differ about the amount of economically recoverable coal still available in the world, but they agree that the quantity is very large: almost certainly above 600 billion tons of coal equivalent, or more than 250 years of coal at current levels of use (for example, Wilson et al., 1980, and Hafele et al., 1981). The total coal resource is believed to total 10,000 billion tons of coal equivalent, or more than 1,200 years of the current annual commercial energy consumption of the entire world (Hafele et al., 1981, pp. 45–47). Moreover, other solid fossil fuels such as oil shale and tar sands are also available in abundance. Besides meeting needs for solid fuels, these resources are clearly sufficient to meet all demands for liquid and gas fuels for a century or more.

Third, the production of synthetic fuels from coal and oil shale is technologically feasible. We know we can make substitute liquid and gas fuels from coal, oil shale, and other solid fossil energy resources. No technological breakthroughs are required. No significant technological uncertainties remain. Any technological improvements will help, of course, but synfuels plants have already been built that work.

Fourth, replacing petroleum and natural gas fuels with synthetic fuels is easy on the end user. For end users, whether industrial plants, automobile owners, or home owners, it will be far simpler to buy and use synthetic liquid and gas fuels than to replace end-use equipment in order to utilize other energy delivery forms, such as electricity. Rather than calling for a great many capital investment decisions in adjusting to rising oil and gas prices, a synfuels strategy adapts the capital stock more centrally. Energy users can continue to buy familiar fuels for familiar equipment, which will be especially important for the transportation sector, where widespread fuel switching would almost certainly be expensive and disruptive.

Fifth, taking the lead in synfuels production will give the host area economic advantages. If synthetic fuels are going to be a part of the world's energy future, the countries and regions who are ready when these options become economically competitive will have important advantages. They will be the exporters of cheaper liquid and gas fuels to other areas sagging under the costs of petroleum and natural gas. To some degree, they may be the new Saudi Arabias and Texases of the post-petroleum era.

Why, then, not go full speed ahead? The case against accelerating the development of synthetic fuels is usually based on some version of five arguments.

First, petroleum and natural gas are readily available, relatively cheap, and likely to remain that way for a considerable time yet. It is hard to get very excited about long-term scarcities of petroleum and natural gas when current markets show signs of a glut in both fuels. Furthermore, many forecasts of world oil prices show little increase, at least in real terms, for the rest of the century; and many observers predict that new natural gas reserves will turn up in deep formations and new parts of the world. Because synthetic fuels are unlikely ever to compete economically with natural liquid and gas fuels, even if the natural fuel recovery costs are relatively high, it is risky to get synfuels ready in twenty years for markets that may not be there until forty.

Second, a sizable increase in the use of coal is hazardous to the environment and human health. We are still learning about impacts of coal use, but we already know that there is plenty to worry about. Coal use is associated with such possible problems as acid precipitation, carbon dioxide levels in the atmosphere, and water pollution. Substituting synthetic fuels for petroleum fuels and natural gas is likely to be socially acceptable only if it can be shown not to contribute to such problems. In addition, some synthetic fuels may present risks to human health (for example, direct coal liquefaction), and most would require mining and waste disposal at scales that would have significant environmental impacts. It is important not to rush too quickly into energy options that may have to be stopped because their impacts are unacceptable.

Third, if and when it makes sense, the private sector will do it. When the best people to produce synthetic fuels, the large energy and chemical corporations, seem so little interested in an option that we know so well, it gives one pause. Several have stuck their toes in the water—such as Exxon, Texaco, and Dow—but none appears to be expanding its efforts very much or advocating public-sector support for a synfuels industry. Maybe this tells us that the time is not yet ripe.

Fourth, the demands on public-sector budgets are many and the fiscal resources are scarce. In the United States, the dominant public policy issue of the later 1980s is likely to be reducing federal budget deficits. Many budget requests will be cut repeatedly, from national defense to social programs. Coming into this setting with requests for funds for an energy-resource/technology option that may be needed in a couple of decades does not offer much hope. Even if it is worth doing, in principle, so are many other things that will not end up with a share of the scarce funds available.

Fifth, solid fossil fuels are energy sources of the past, not of the future. In the high-technology world of the 1980s, it is easier to get enthusiastic about moving in new directions—such as energy from the atom, solar energy, or genetic engineering of biomass resources—than working with bulky, dirty solid fossil fuels, especially when most visions of a synfuels future are so close to technologies already available and when we know that the long-term future will belong to renewable resources. Unlike other long-term energy options, in a sense synfuels are a relatively low-tech expedient for a century or so, not a chance to shape a millennium.

In assessing these pros and cons, one balances two kinds of concerns against each other. First, what are likely to be the costs of acting? What is the chance that we will invest precious public- and private-sector fiscal and managerial resources in an energy option that will never be needed, or at least not needed for a long time after it is ready? What is the likelihood that the development process itself will have undesirable environmental and health impacts? Is an effort to push synthetic fuels likely to be devisive, socially and politically? Second, what are likely to be the costs of failing to act? What is the chance that public policy inattention to synthetic fuels will leave us without a needed alternative when oil and gas prices begin their final climb? Is it possible that the economic and political strains associated with rising prices will threaten our national security and even bring some of our democratic institutions into question?

In spite of the wishes, efforts, and occasional claims of analysts, we have found in the past decade that any process of balancing these concerns involves too many uncertainties for science to produce a conclusive answer. One can argue, for instance, that the costs of failing to act when action is needed are far more draconian than the costs of acting when action is not needed, but one can also argue that the former situation is far less likely than the latter. This makes what would have been a very political process in any event even more political, in the sense that decisions will be determined by the interplay of vested interests and their levels of influence and control.

From this perspective, who has a vested interest in accelerating the development of synthetic fuels in the United States? Regions with coal or oil-shale resources have a vested interest if they are willing to accept the socioeconomic impacts of development, the eastern U.S. coal regions almost certainly, the western regions perhaps. Other vested interests include technology developers who hold patent rights and licenses; liquid and gas fuel distributors and those who produce equipment that uses these fuels, if mid-term shortages of natural liquid and gas fuels would encourage fuel switching; end users if such switches would be painful and expensive, both users of fuels and hydrocarbon feedstocks; and a public policy-making system that is expected to protect the country against such situations, including national security implications of import dependence or fuel scarcity. Vested interests might even include large energy corporations who face public suspicion and criticism as prices rise.

Except for prodevelopment coal regions and technology patent holders, however, the constituency for accelerated development is indubitably iffy, based either on pessimism about the long term or on statesmanship with respect to preparing for it. In effect, it advocates spending now to realize benefits (which admittedly may be quite large) in a future that we have discovered is farther away (maybe much farther away) than the planning horizons of profit-making firms or public agencies in the United States. Most of the benefits would accrue to general rather than particular entities: the public, the industry, or the region.

Who has a vested interest in opposing an acceleration of synfuels development in the United States? The energy companies have a vested interest if they are expected to pay very much of the bill themselves, because the likely payback period is far too long and, besides, their more immediate need is to find more customers for oil and natural gas, not more substitutes for it. Other vested interests include those who believe that a major increase in coal utilization would be harmful to environmental quality and human health, those throughout the political spectrum who would prefer that public funds be allocated for other purposes, and those who by reason of philosophy or vested interest prefer other energy alternatives for the long term.

Compared with the constituency in favor of accelerated development, this latter constituency is less iffy and more focused on particular rather than general vested interests. Is it not at all surprising, then, that when what had seemed to be an imminent danger of national energy insecurity receded, the new consensus agreed on the demise of synthetic fuels as a public policy priority? Only a widespread sense of threat, difficult to sustain over a long period, had stimulated the provisions of the Energy Security Act in the first place, and even then many supporters of synfuels had suggested that the wrong path had been taken.

All along, in fact, there had been voices calling for attention to some kind of middle ground. Wasn't it possible to resolve to accelerate the learning about synthetic fuels without resolving to accelerate their commercial production? Couldn't we "buy insurance" against an uncertain future by investing a relatively modest amount of public money in the guinea-pig stage, along with technology improvements, so that we could move to large-scale commercial production more quickly if and when we decided we would need it a decade or so in the future? But this view never spread very far beyond the energy policy research community itself and, like many attempts to travel the middle of the road in a rather polarized policy arena, never seemed to appeal very much to either of the more polar constituencies.

PLAUSIBLE ALTERNATIVE FUTURES FOR SYNTHETIC FUELS

Given the environment for decision making summarized above, let us consider how the evolving environment for public policy making might affect the prospects of synthetic fuels in the United States. Two very different kinds of environments can be posited to bound a universe of possible futures: one in which government takes no action whatsoever to encourage synthetic fuels in the next several decades and one in which government acts resolutely for several decades to bring about a commercial synthetic fuels industry as

Table 10–1
Two Baseline Scenarios for Synthetic Fuels Development in the United States: Likely Production Levels in Barrels of Oil Equivalent per Year for Liquid/Gas Fuel Markets

	Business as Usual	Synfuel Mobilization
2005	20,000 - 50,000	250,000 - 300,000
2015	100,000 - 200,000	1 - 2 million
2025	500,000 - 1,000,000	4 - 6 million

quickly as technologically and institutionally feasible. We will call these the "business as usual" case and the "synfuel mobilization" case (Table 10–1).

Prospects for Synfuels with Business as Usual

In the 1980s business as usual means that synthetic fuels must be developed by the private sector without government help, in a time when natural oil and gas fuels are available at relatively low prices. In other words, for development to take place, private firms must conclude that synfuels can be produced and delivered more cheaply than inexpensive alternatives in the relatively near future, under realistic market conditions.

Let us assume that world oil prices will follow a path roughly approximated by current middle-of-the-road projections (Curlee, 1985): staying below $20 per barrel in mid–1980s prices until at least the early 1990s, then rising during the late 1990s to a level of $30–35 per barrel. Prices during this period are more likely to oscillate around the general trend than to change smoothly (Curlee, Reister, and Fulkerson, work in progress), which will add to perceptions of uncertainty in forecasting the future as a basis for investment decisions.

Given this kind of environment, current patterns of activity indicate that synfuels investments will be limited in the near future to two kinds of activities: occasional conversions of coal to hydrocarbon feedstocks (even the rather isolated but important cases of Tennessee Eastman and Dow Chemical represent investment decisions of the late 1970s, not the mid–1980s); and a few uses of gasification as a coal-cleaning approach for electric power generation, along the lines of the Cool Water plant in California and a facility proposed by Potomac Electric Power in Maryland. The next new facilities for synthetic fuels production, at least for liquid and gas fuel markets in the United States, will await a perception that petroleum prices are likely to reach unprecedentedly high levels within less than two decades—on the order of

$50 per barrel of crude oil in mid–1980s dollars—and stay at those levels or higher. This suggests construction of the first such pioneer facilities beginning no earlier than the late 1990s, following go-ahead decisions by industry in the mid–1990s. Assuming 5–7 years for construction and shakedown for large pilot plants and several years of operation to provide baseline experience, plus 8–10 years for siting, permits, and construction of the first generation of full-scale commercial plants (with the industry still cautious at the outset), this will mean synfuels production levels no higher than 20,000–50,000 barrels of crude oil equivalent (bcoe) by 2005, 100,000–200,000 bcoe by 2015, and 500,000–1,000,000 bcoe by 2025—less than half the 1990 goal of the Energy Security Act.

Prospects for Synfuels with Maximum Government Support

The Energy Security Act grew out of a short-lived fascination with the idea of a wartimelike mobilization in the interest of energy security in the United States: reducing our dependence on oil imports by a national commitment to develop substitutes from domestic resources, equivalent to our effort in the 1960s to place people on the moon. What if, somehow, we were able to renew this commitment and continue it? It is hard to imagine that such a change in policy direction could occur before 1989 and that industry confidence could be restored in less than several years after that. Perhaps industry might be induced to start construction of a half-dozen plants in a 10,000–50,000 bcoe size range by 1992–1993, beginning operation by 1997–2000. With strong continuing public support an equal number of similar facilities might be under construction before the pioneer plants start production. With the success of pioneer plants and continued public support a full commercial industry could be under way by the middle of the first decade of the next century, reaching production levels in the range of 250,000–300,000 bcoe by 2005, 1–2 million bcoe by 2015, and 4–6 million bcoe by 2025. Higher production levels would be possible, but the aggregate requirements for sites, capital, materials, and other key ingredients make them unlikely.

Possible Departures from Business as Usual

This kind of continued unrestrained government support is hard to imagine, given the business as usual scenario of world oil prices, but it is more instructive to consider what might change business as usual. Table 10–2 identifies the two most likely changes: changes in the world oil market, either increasing or decreasing the political and economic attractiveness of investments in synfuels, and changes in our understanding of the impacts of increased coal use, either strengthening or moderating our collective concerns about its acceptability. Other possible changes include technological breakthroughs making synfuels more competitive economically and the appearance

Table 10–2
Four Scenarios for Changes in Business as Usual for Synthetic Fuels Development in the United States: Likely Production Levels in Barrels of Oil Equivalent per Year for Liquid/Gas Fuel Markets

		Acceptability of Coal (compared with Business as Usual)	
		Higher	Lower
Oil Prices (compared with Business as Usual)	Higher	2005 : 50,000-100,000 2015 : 500,000 2025 : 2-3 million	2005 : 50,000 2015 : 200,000 2025 : 500,000
	Lower	2005 : 10,000-20,000 2015 : 50,000-100,000 2025 : 0.5-2 million	2005 : 0 2015 : Negligible 2025 : Negligible

of political leadership to forge a consensus in favor of some level of government support for synfuels development that is less than a hypothetical maxim. Consider the four change scenarios identified in Table 10–2.

Oil prices higher, reduced concerns about the acceptability of coal. This is the most encouraging scenario for the prospects of synthetic fuels. If both of these changes were to occur, synfuels would be more likely to get public policy support and to be judged a good business investment. If the changes occurred by the early 1990s, several pilot ventures would probably appear by the mid–1990s, beginning operation early in the 2000–2005 period. If both changes were to continue and strengthen, and the experience with the pilot plants were positive, commercial investments would follow, but at a slower pace than in the mobilization case because of higher business risks. Production levels might reach 50,000–100,000 bcoe by 2005, as much as 500,000 bcoe by 2015, and 2–3 million bcoe by 2025.

Oil prices higher, increased concerns about the acceptability of coal. Here the winner is more likely to be fuel switching than synthetic fuels, as the public policy focus shifts toward R&D support and incentives for alternatives other than solid fossil-based fuels. Given sufficiently sharp price increases, some attention would be given to synfuels as one of many options in a diverse portfolio, but they would generally be kept in reserve in case other options do not meet energy needs. Likely production levels for fuels, probably less if attractive alternatives are found in the meantime, would be in the range of 50,000 bcoe by 2005, 200,000 bcoe by 2015, and 500,000 by 2025.

Oil prices lower, reduced concerns about the acceptability of coal. This scenario is the most likely for R&D to improve the prospects of synthetic fuels: a positive attitude toward coal, plus time for new ideas to be generated and converted into cheaper and cleaner production technologies. Production would grow more slowly, but it might well be on the upswing by the second quarter of the century: perhaps 10,000–20,000 bcoe by 2005, 50,000–100,000 bcoe by 2015, and 0.5–2 million bcoe and rising by 2025. Whether very much of the R&D would be carried out in the United States, however, is a different question.

Oil prices lower, increased concerns about the acceptability of coal. Lastly, this combination of circumstances would effectively kill synthetic fuels in the United States, except as a minor area of R&D. U.S. energy institutions would continue to monitor and perhaps participate in technology-application experience in other countries, but domestic production would be negligible, at least for liquid and gas fuel markets.

A SUMMARY OF THE PROSPECTS FOR SYNFUELS

Barring some other change, such as political leadership or a technological breakthrough, the prospects add up as follows. Synthetic fuels face an uncertain future in the United States. In order for the private and public sectors to make a joint commitment to such an economically risky way to meet future energy requirements, the nation will have to be convinced of a clear need decades into the future. But projecting energy conditions even a few years away has become an exceedingly shaky undertaking.

Uncertainties, in fact, are pervasive. Leading the list is the continuing uncertainty about world oil and gas reserves, including potentials for enhanced recovery. Oil and natural gas will have to become scarce indeed before synthetic fuels are cheaper in the marketplace. Other uncertainties include the public acceptability of coal use, the availability of public-sector financial support, trends in the liquid and gas fuel needs of developing countries, possible effects of political agendas and events, potentials for sharp reductions in the costs of synthetic fuels technologies, and potentials for breakthroughs in other energy technologies, particularly those based on renewable energy sources.

These uncertainties get in the way of decisions, public or private, to move ahead with synthetic fuels. Based on the scenarios outlined above, consider what a synthetic fuels industry would need in order to develop significant momentum in the United States by the first decade of the next century. An assortment of pioneer commercial plants would need to be built and operating by the end of this century, not necessarily in the United States alone, but under conditions that build an institutional capability in the United States and provide credible, validated information about economic, technological, and impact characteristics of synfuels technologies. At a minimum, this would require a major upturn in oil prices by the early 1990s, plus a broad consensus in the United States by the end of that decade that a sizable increase in coal development and use is not only acceptable but desirable.

What seems more likely is that the United States will make very little progress with synthetic fuels during the remainder of the 1980s, except for scattered experience from a few isolated ventures undertaken mainly by corporations needing hydrocarbon feedstocks, who see a combination of a shorter-term economic break-even situation and a hint of a chance at longer-term technological leadership, or by utilities and their partners who utilize coal gasification in power generation at a relatively small scale. Generally, public money will be scarce, and market conditions will be unattractive to private investors in the production of fuels. During the 1990s concerns about energy scarcity and security are likely to return, and they will mean renewed interest in synthetic fuels, even a few calls for mobilization. But our national history of frequent stops and starts will make decision makers understandably cautious, and the leadership in technology demonstration may well be taken by other countries. The critical period will probably be the early part of the next century, as we get closer to the need to switch from oil and gas and see our future energy needs more clearly, including, for instance, the outlines of our next era of transportation systems, the place of hydrocarbon feedstocks in markets for materials, and the potentials for end uses of electricity.

At that point, we may find that we are too late. The liquid and gas hydrocarbon fuel crunch will be upon us, with too little time to get alternatives ready. On the other hand, we may find that, with a certain amount of awkwardness in the transition, we can get along without synthetic fuels just fine in weaning ourselves away from depletable fossil energy sources. This uncertainty brings us back to the fundamental predicaments that have brought our national synthetic fuels program virtually to a halt. First, unless the eventual market for synfuels will be quite large, the interest of the private sector will be rather small; but if the eventual level of production will be quite large, the management challenge and the impact/acceptability issues will also be quite large, which presents industry with considerable risks given the lack of a grandfather clause in federal government commitments. Second, it requires longer to build a commercial-scale synfuels facility than the investment time horizons of major private-sector decision makers; this means that a decision to proceed at this time will require government initiative—when the payoff is also farther away than the election-oriented time horizons of major public-sector decision makers.

IMPLICATIONS AND OPTIONS

The main implication of this project is that we are taking a chance. The late 1970s gave us a hint of the impacts on our national economy and psyche of a perceived scarcity in petroleum fuels. We know that we face a much larger sustained transition away from petroleum and natural gas fuels, almost certainly within the lifetimes of some individuals now in decision-making positions. We know that the transition will be especially difficult for the transportation sector, which will be seriously affected by a steep increase in the price of hydrocarbon liquid fuels and which will probably find a shift to other energy sources problematic.

Taking no public policy actions to accelerate the development of a synfuels industry beyond a business as usual pace means we are betting that we can cope with the transition by means of quick ad hoc responses or else that the transition is far enough away that policy actions can be deferred indefinitely. That we will be able to cope is certainly possible. For example, we may have at least two capabilities to fall back upon: international capabilities and limited domestic private-actor capabilities.

Other countries will continue to be interested in synthetic fuels even if we are not. There, too, the main commercial thrust for the next decade is likely to be hydrocarbon feedstocks for industrial production rather than fuels to meet energy needs, for example, in Sweden, Greece, and Finland, but a number of countries are proceeding either with commercial synfuels plants or with active technology-demonstration programs. Besides South Africa, which is obviously the special case, leading countries will include China, West Germany, Japan, Great Britain (an underground coal gasification pilot project is especially notable), and perhaps the USSR and Canada. In addition, the large-scale production and use of alcohol fuels in Brazil is a source of valuable lessons about introducing synfuels into an economy. Moreover, experiments with the utilization of oil shale are likely to continue in such areas as the Near East. As a result, even if synfuels technology development and application does not thrive, it will not die. The experience of others with synthetic fuels, like the experience of France with breeder reactors, adds capabilities that might eventually be useful to many others.

Meanwhile, firms based in the United States will not be completely out of the business. Tennessee Eastman's use of the Texaco coal gasification process to produce acetic anhydride and the Cool Water Coal Gasification plant in California, which supports a 100 mwe electric power plant, both of which are economically competitive under current economic conditions, will lead a few others to try similar technology applications, and Texaco gasifiers have proved to be competitive elsewhere, for example, in Sweden. It appears that the Great Plains Coal Gasification Plant in North Dakota will continue to operate unless the contracted buyers of the product gas can find ways to back out of their obligations, and there is a small chance that the shale-oil project of Union Oil may survive with last-gasp support from the dying Synthetic Fuels Corporation. Technology R&D has been scaled back considerably in the 1980s, but it continues with some cost sharing by the Department of Energy, oriented toward such possibilities as improved gasification processes and "low-severity" processes for liquefaction such as lower-temperature pyrolysis. Moreover, it is hard to imagine that U.S. synfuels experts will not continue to serve as consultants for ventures in other countries and that major U.S. architect-engineering firms will not find a way to be involved in some of the new plant construction wherever it occurs.

We cannot, however, depend on this to be sufficient. The problem is that—given a history of uncertainty, irresolution, fluctuations in world oil prices, and inconsistent government policy—our foresight is likely to be limited. Because some of the processes and events that will push oil prices to levels that economic and social systems cannot handle without unacceptable pain will have been political, and because many of the responses that will shape both markets and public policy actions will have their bases in psychology rather than resource economics, many energy experts will be surprised; and a policy-making system that has had too many other crises to resolve to respond to conflicting signals from the energy experts will probably be unprepared.

Clearly, it would be misleading to imply that the day will arrive when a shift to alternative energy sources will be absolutely necessary, beyond question. But a time will arrive, perhaps a period of a few years, when the economic and national security costs of relying on petroleum and natural gas in historic quantities will become unacceptable, with little prospect that the situation will change dramatically. Unless other options are in place by then—some combination of efficiency improvements, fuel switching, and synthetic fuels production—the recriminations will be intense and the risk of hasty, unwise policy actions will be considerable.

If a part of the current answer will have to be synthetic fuels use, possibly for the transportation sector, then the preparation will have to begin at least a decade or more earlier: sustained action to build and operate an assortment of synfuels plants at scales, in locations, and under conditions of information production and access that build the basis of experience that we will need. If the preparation has not begun soon enough, we will find ourselves under heavy pressure to do something in a hurry, looking at an assortment of options mainly developed outside the United States, tested under conditions that reflect different standards from ours, characterized by information whose validity is open to question, and raising issues to be resolved by decision-making structures that have often foundered under the stresses of satisfying diverse interests in a hurry.

Meanwhile, one can imagine, protests will grow. Farmers, truckers, airlines, and commuters will see their costs climb. Tourist destinations will be affected by costs and scarcities in transportation fuels. Nationalists will be concerned about increased dependence on foreign technology, facility management, and flows of profits out of the country. At the same time, concerns about environmental and health risks, along with formidable requirements for assembling resources and managing major new operations, will lead to delays with new synfuels projects, while climbing interest rates associated with rising energy prices, risks of economic turbulence, and competition for scarce capital add to facility costs. The strains on our political and economic fabric could be substantial: consider 1973–1974 and 1979 sustained and multiplied.

What can we do to avoid these consequences? We know there are some things we cannot do. We cannot create and sustain large new public institutions. We cannot spend large amounts of public money. We cannot convince private firms to move very quickly with synthetic fuels on their own. We may, however, be able to find ways to buy some inexpensive insurance by returning to the ideas associated with the middle ground of the synfuels policy debate of the late 1970s that reflected a broad consensus in the professional energy community even if it had little policy impact at the time. Refined a little by what we have learned since then, along with the federal budget realities of the 1980s and 1990s, this consensus suggests the following policy directions, in terms of both technology and institutional improvements.

Improvements are needed in coal technology. The 1970s showed us how little we really know about coal as a resource and as a hydrocarbon building block. Because coal was first too familiar and then too undesirable, we had neglected it as a focus of research, letting its science and technology base fall behind those of our other major energy options. With coal and other solid fossil fuels potentially so important for the long term, our program of energy research needs to include continued attention to these resources, ranging from geology to chemistry and the health sciences.

Improvements are also needed in synfuels technology R&D. Along the lines of the Department of Energy's current clean coal R&D program, but larger, we need to reestablish a public-sector/private-sector partnership through cost-sharing projects that show a stable public policy commitment to include improved U.S. synfuels technologies in our energy portfolio for an uncertain future. The emphasis should be on "proof of concept" for technologies that add resiliency to the portfolio, particularly technologies for producing transportation fuels and industrial feedstocks.

The federal government needs to go beyond technology R&D to assure two kinds of benefits from experience with operational synfuels facilities without repeating the expensive Great Plains experiment: building synfuels production capability rather than production capacity within U.S. private-sector institutions and generating credible information about facility performance and impacts. The best prospects are to encourage coal gasification for electric power generation and to support U.S. firms active in synfuels development in other countries; but relatively low-cost incentives for other domestic facilities, such as in the chemical industry, are also worth considering.

Early in the 1980s a Department of Energy Energy Research Advisory Board report noted that "our energy system is a web of machines and people;

the current budget is focused on machines" alone. Others have pointed out an apparent inconsistency in our energy R&D policy: for a decade most people have agreed that our main energy challenges are not primarily technological, but have involved a search for new options and capabilities focused almost exclusively on technology (for example, Wilbanks, 1982). To assure that our economic and social options as a nation will not be eventually limited by our energy options, a part of our strategy should be to improve the institutional setting for action when and if action is needed.

Toward that end, we can, first of all, create more effective mechanisms for private-sector/public-sector cooperation. Without compromising the public interest, we need more innovative policies to assist U.S. firms interested in maintaining synfuels capabilities, to get public access to information from private firm experience, and to prepare contingency plans for the "surge deployment" (Harlan, 1982) of synfuels facilities if, twenty years or so from now, we decide we will need them. Recent policy initiatives to increase technology transfer from the public to the private sector and to improve the competitiveness of U.S. firms in international markets are promising moves in this direction.

Second, we can broaden the consensus about the pros and cons of synfuels development. Through a combination of continuing R&D programs to clarify environmental and health implications of synfuels options and active programs of information dissemination inviting questions from any observers who may be skeptical, the federal government can build from the consensus that had begun to emerge by 1980 toward a much broader agreement in U.S. society about the role of synthetic fuels: how, when, and under what conditions to accelerate their use. Some political leadership will be required, but the economic and political costs can be relatively small.

Third, we can develop more effective mechanisms for resolving controversies about large-scale synfuels production. We need better ways to resolve disputes than heavy-handed federal government directives on the one hand or slow and erratic legal processes on the other. Given a decade to prepare, this is a question for research and development now, a salient example of the promise of "non-hardware R&D" (Kash et al., 1976; Landsberg et al., 1979) to contribute to the solution of energy problems. Possibilities that deserve systematic attention include innovative approaches to assuring the consent of parties at risk (for example, Wilbanks, 1981), ways to build better communication between groups who often see each other as adversaries (for example, Murray, 1978), and mechanisms for identifying social concerns and incorporating them into technology R&D priorities.

Such a strategy would cost only a fraction of a Synthetic Fuels Corporation, but it would put us in a position to react much more quickly in the future than we could right now to an imminent petroleum/natural gas shortage. To do much less is to abdicate the federal government's responsibility for national preparedness.

THE BIGGER ISSUE

The synthetic fuels issue is interesting for its own sake, because of the long-term stakes involved; but it is even more interesting as an example of a larger issue: whether or how a democratic society composed mainly of parties with short-term time horizons can in fact prepare itself for the long term. The actual issue, of course, is more subtle. Public support for basic scientific research with all of its connections to economic competitiveness and higher education is as well established in the United States as motherhood and apple pie, although the rising cost of scientific equipment leads actions to fall increasingly short of intentions. The real issue is not with science for the long term but with technology for the long term, especially when technology needs and applications are likely to be large in scale and perceived as risky by many in the democratic society.

It can be argued that in this respect we face two formidable challenges in our society. The first is that our key decision makers face powerful incentives to focus on the near term rather than the long term. Whether they are corporate executives measured by their accomplishments during 2–3 years or government officials concerned about the next election, worrying about the long term is a luxury or, at best, a commitment that must be hidden or compromised or sold in terms of nearer-term benefits. The second is that substantial parts of society have become convinced that we can avoid significant risks or, at least, that risks should not be taken by the few on behalf of the many, and risks are far more perceptible from action than inaction. Combined with the very real uncertainties about long-term conditions and needs, these challenges have largely put technology R&D in the United States in neutral whenever policy involves time horizons beyond 5–7 years and technology applications appear both large and potentially risky. Very simply, from the points of view of decision makers, there are more promising things to spend time and money on.

This situation, however, leaves us vulnerable to a host of futures that are possible but relatively unlikely, such as a problem with carbon dioxide accumulation in the earth's atmosphere; and it leaves us at a disadvantage in world markets beyond a period of 5–7 years compared with others who have paid more attention to that time horizon, investing in technology options that promise to be better even if their payoff periods are longer.

In particular, a focus on the near term, plus support for very long-term basic research, makes it too easy for us to neglect the period in between: that time period 5–7 to 20–25 years away for which public-sector/private-sector cooperation is most likely to be important. This is the period in which social needs must be met largely by technologies we already know about, incrementally improved and fine-tuned. It is the period that will develop the institutional framework for responding to longer-term needs. And it is exceedingly unclear who is taking the responsibility for this period. To begin to correct such a situation, we need more public attention to several national needs.

First, improved mechanisms for consensus building about technology developments and applications are needed. It has been suggested elsewhere that American society has changed from one in which President Eisenhower felt it necessary to warn of the unbridled power of the "military-industrial complex" to one in which a wide range of individuals and groups now participates in most important decisions. Unfortunately, however, our policy-making structures have yet to catch up with our changed society (Branscomb, 1978); meanwhile, "adversarial decisionmaking structures convert uncertainties into disagreements or, worse, antagonisms, and prospects for action fade away time after time" (Wilbanks, 1981, p. 1). Clearly, the problems are worst for large-scale facilities, where the visibility of a technology application is higher and the possible impacts are larger (Wilbanks, 1984; Wilbanks and Lee, 1985); but the growing size of our population and our economy will continue to require large-scale technology applications, even if more needs can be met by smaller-scale technologies. Here, the synthetic fuels experience indicates that we need a national strategy that includes at least three elements: R&D attention to "nonhardware" issues, the use of current technology-application issues to test new ideas about consensus-building approaches (for example, in hazardous waste disposal), and a deliberate effort to incorporate social preferences into technology R&D planning, at least by the public sector.

Second, improved mechanisms are needed for public-sector/private-sector cooperation in technology development for the midterm. A persistent legacy of perceived and, in some cases, perhaps real excesses of big business is an expectation on the part of many that the electoral process, legislative action, and judicial procedures will provide checks and balances in the interest of those otherwise without enough economic influence to participate in decisions. The adversarial processes that have developed as a result have often led to decisions about what not to do rather than agreements about what we should do, and they have inflated the importance of the courts in resolving disagreements. As a consequence, instead of cooperating to extend time horizons and improve competitiveness internationally we are too often wrapped up in our own parochial infighting. In the meantime, countries where the public and private sectors are able to work as closer partners are beating us in the world marketplace, undermining employment and real incomes in the United States, and our competitiveness is in danger of declining further as we move into a mid-term time period for which our own short-term time horizons have prepared us so poorly. Synthetic fuels are by no means the most important example of this problem.

Third, technology R&D needs a more stable environment. Our history since the mid–1970s of public policy toward fossil energy R&D is a classic case of irresolution in technology R&D support, but it is not the only case. Whenever a field of R&D moves out of the province of the research community into the public spotlight, it seems to be subject to changes in direction that are so frequent that little R&D progress can be made (Kash and Rycroft, 1984). A more wholehearted endorsement by the federal government of multiyear R&D funding would be a positive step; but the more fundamental answer is to base technology R&D priorities on widely shared national perspectives about problems that need technology solutions, especially solutions beyond the time horizons of private sector R&D. This, in turn, adds to the importance of building consensus across boundaries between parties at interest.

Fourth, increased R&D attention needs to be given to the human dimensions of technology deployment. As important as any of the needs is recognizing that technologies are deployed by and for people—and that the human dimensions of technology development and use are researchable themselves, not merely the realm of untidy political processes. Whether the issue is social preferences, management effectiveness, or conflict resolution, effective research tools and skills are available to help. Too often, technology R&D has been focused on technological fixes to some problems and "nonhardware" R&D focused on social fixes to others, when both kinds of R&D can contribute to strengthening both kinds of fixes to the same problems (for example, conflict resolution with respect to technology deployment and data-system technologies with respect to the performance of social institutions). It is time to recognize the rich multidisciplinary nature of virtually every national issue in this world that is so technological yet so human and to give equivalent R&D attention to both hardware and nonhardware dimensions of the issues.

Clearly, these points pose questions without providing answers; significant improvements in our prescience with respect to technology development and deployment will not come easily. Just as clearly, the major contributor to such improvements will have to be political and business leadership, because time horizons at the grass roots are as limited as in our larger institutions. But the issues will arise time and time again—in national security, bioengineering, communication and privacy, automation and jobs, waste disposal, health care, environmental protection, and many other fields of technology application. Given the uncertain nature of our future, combining a well-informed public with the idea of buying relatively inexpensive insurance against undesirable contingencies may point us in the right directions, even if it does not smooth all the pathways.

REFERENCES

Branscomb, L. February 1978. Testimony before a joint hearing of the U.S. Senate Committee on Commerce, Science, and Transportation and the U.S. House Committee on Science and Technology.

Curlee, T. R. 1985. "Forecasting World Oil Prices: The Evolution of Modeling Methodologies and Summary of Recent Forecasts." *Materials and Society* 9:413–42.

Curlee, T. R.; Reister, D.; and Fulkerson, W. Work in progress. Energy Division, Oak Ridge National Laboratory, Oak Ridge, Tenn.

Hafele, W., et al. 1981. *Energy in a Finite World*. Cambridge, Mass.: Ballinger.

Harlan, J. K. 1982. *Starting with Synfuels*. Cambridge, Mass.: Ballinger.

Kash, D. E. et al. 1976. *Our Energy Future: The Role of Research, Development, and Demonstration in Reaching a National Consensus on Energy Supply*. Norman: University of Oklahoma Press.

Kash, D. E., and Rycroft, R. W. 1984. *U.S. Energy Policy: Crisis and Complacency*. Norman: University of Oklahoma Press.

Landsberg, H. Winter 1986. "The Death of Synfuels." *Resources, Resources for the Future*, no. 82:7–8.

———. 1979. *Energy: The Next Twenty Years*. Cambridge, Mass.: Ballinger.

Murray, F. X., ed. 1978. *Where We Agree*. Report of the National Coal Policy Project, 2 vols. Boulder, Colo.: Westview Press.

Wilbanks, T. J. September 1981. "Building a Consensus about Energy Technologies." ORNL 5784. Oak Ridge National Laboratory, Oak Ridge, Tenn.

———. January 1982. "Some Issues in an Energy R&D Strategy for the Long Term." A paper delivered at the Symposium on Energy R&D Strategies and National Energy Policy, American Association for the Advancement of Science, Washington, D.C.

———. 1984. "Scale and the Acceptability of Nuclear Energy." In *Nuclear Power: Assessing and Managing Hazardous Technology*, ed. by M. Pasqualetti and D. Pijawka. Boulder, Colo.: Westview Press, pp. 9–50.

Wilbanks, T. J., and Lee, R. 1985. Policy Analysis in Theory and Practice." In *Large-Scale Energy Projects: Assessment of Regional Consequences*, ed. by T. R. Lakshmanan and B. Johansson. Amsterdam: North-Holland, pp. 273–303.

Wilson, C. L. et al. 1980. *Coal: Bridge to the Future*. Report of the World Coal Study. Cambridge, Mass.: Ballinger.

Selected Bibliography

AME Technology, Inc. 1976. *Social, Economic, and Environmental Aspects of Coal Gasification and Liquefaction Plants: Final Report*. Lexington, Ky.: Office of Research and Engineering Services, College of Engineering, University of Kentucky.

Ausness, Richard C. 1976. *Legal Institutions for the Allocation of Water and Their Impact on Coal Conversion Operations in Kentucky*. Report no. 95. Lexington, Ky.: University of Kentucky Water Resources Research Institute.

Borkin, Joseph. 1978. *The Crime and Punishment of I. G. Farben*. New York: Free Press.

Bruder, G. C., and Gentile, C. L. 1981. "State Synfuel Agencies' Role in the Federal Program." *State Government*. 54:14–20.

Calzonetti, Frank J. 1981. *Finding a Place for Energy: Siting Coal Conversion Facilities*. Washington, D.C.: Association of American Geographers.

Committee for Economic Development. 1979. *Helping Assure Our Energy Future: A Program for Developing Synthetic Fuel Plants Now*. New York: Committee for Economic Development.

Congressional Research Service. 1980. *Synfuels from Coal and the National Synfuels Production Program: Technical, Environmental, and Economic Aspects*. Washington, D.C.: Congressional Research Service.

———. March 1981. *The Pros and Cons of a Crash Program to Commercialize Synfuels*. Washington, D.C.: Congressional Research Service.

Cowser, K. E., and Richmond, C. R., eds. 1984. *Synthetic Fossil Fuel Technologies*. Stoneham, N. H.: Butterworth's.

Crow, Michael M., and Hager, G. L. August 1985. "Political Versus Technical Risk Reduction and the Failure of U.S. Synthetic Fuel Development Efforts." *Policy Studies Review* 5(1):145–52.

Doerell, Peter E., ed. 1982. *World Synfuels Project Report*. San Francisco: Miller Freeman Publications.

Dow, Alan. 1981. *Synthetic Fuels from Coal and Oil Shale*. Lexington, Ky.: Appalachia-Science in the Public Interest.

Ellington, Rex T. 1977. *Liquid Fuels from Coal*. New York: Academic Press.

Goodwyn, Craufurd, ed. 1981. *Energy Policy in Perspective: Today's Problems, Yesterday's Solutions*. Washington D.C.: Brookings Institution.

Hammond, Ogden, and Zimmerman, Martin. July/August 1975. "The Economics of Coal-Based Synthetic Gas." *Technology Review* 77:42–51.

Harlan, J. K. *Starting with Synfuels*. 1982. Cambridge, Mass.: Ballinger.

Hederman, W. F. 1978. *Prospects for the Commercialization of High-Btu Coal Gasification*. Santa Monica, Calif.: Rand Corporation.

Hill, Richard F. 1980. *Synfuels Industry Development*. Washington, D.C.: Government Institutes.

Hoffman, E. J. 1982. *Synfuels: The Problems and the Promise*. Laramie, Wyo.: Energon Company Publications.

Horwitch, Mel. 1980. "Uncontrolled and Unfocused Growth: The U.S. Supersonic Transport (SST) and the Attempt to Synthesize Fuels from Coal." *Interdisciplinary Science Reviews* 5(3):231–44.

Hundeman, Audrey S. 1980. *Pollution and Environmental Aspects of Fuel Conversion*. Springfield, Va.: National Technical Information Service.

Hunt, Daniel. 1983. *Synfuels Handbook*. New York: Industrial Press.

International Energy Agency. 1982. *Coal Liquefaction: A Technology Review*. Paris: Organization for Economic Cooperation and Development.

Kash, D. E., and Rycroft, R. W. 1984. *U.S. Energy Policy: Crisis and Complacency*. Norman: University of Oklahoma Press.

Krammer, A. January 1981. "Technology Transfer as War Booty: The U.S. Technical Oil Mission to Europe, 1945." *Technology and Culture* 22:68–103.

Lambright, W. H.; Crow, M. M.; and Shangraw, R. May 1984. "National Projects in Civilian Technology." *Policy Studies Review* 3(3–4):453–59.

Lash, J., and King, L. eds. 1983.*The Synfuels Manual: A Guide for Concerned Citizens*. Washington, D.C.: Natural Resources Defense Council.

Lee, Bernard S. 1982. *Synfuels from Coal*. New York: Aiche.

Lefevre, Stephen R. 1984. "Using Demonstration Projects to Advance Innovations in Energy and Communications." *Public Administration Review* 44:483–90.

Leistritz, F. Larry, and Murdock, Steve H. 1981. *The Socioeconomic Impact of Resource Development: Methods for Assessment*. Boulder, Colo.: Westview Press.

Murphy and Williams, Consultants. 1978. *Socioeconomic Impact Assessment: A Methodology Applied to Synthetic Fuels*. Washington, D.C.: U.S. Department of Energy.

Myers, C. W., and Arguden, R. Y. January 1984. *Capturing Pioneer Plant Experience: Implications for Synfuel Projects*. Report N–2063–SFC. Santa Monica, Calif.: Rand Corporation.

Nash, Gerald. 1968. *United States Oil Policy, 1890–1964*. Pittsburgh: University of Pittsburgh Press.

National Research Council. 1977. *Assessment of Low and Intermediate BTU Gasification of Coal*. Washington, D.C.: National Academy of Sciences Press.

———. 1977. *Assessment of Technology for the Liquefaction of Coal*. Washington, D.C.: National Academy of Sciences Press.

———. 1980. *Refining Synthetic Liquids from Coal and Shale: Final Report of the Panel on R&D Needs in Refining of Coal and Shale Liquids*. Washington, D.C.: National Academy Press.

Nowacki, Perry. *Coal Liquefaction Processes*. 1979. Park Ridge, N.J.: Noyes Publications.

———. 1980. *Health Hazards and Pollution Control in Synthetic Liquid Fuels Conversion*. Park Ridge, N.J.: Noyes Publications.

Pelofsky, Arnold H., ed. 1977. *Synthetic Fuels Processing: Comparative Economics*. New York: Marcel Dekker.

Rosenbaum, Walter. 1980. "Notes from No Man's Land." In *Environment, Energy, and Public Policy*, ed. by Regina Axelrod. Lexington, Mass.: Lexington Books, pp. 61–79.

Schwaderer, Roseann. 1980. *Synfuels Handbook*. Washington, D.C.: Coal Week.

Seidman, David. 1980. *Values in Conflict: Design Considerations for a Two-Stage Synfuels Development Strategy*. Santa Monica, Calif.: Rand Corporation.

Steele, Henry B. 1979. *Economic Potentialities of Synthetic Liquid Fuels from Oil Shale*. Salem, N.H.: Ayer Company Publications, Inc.

Stenehjem, E. J., and Allen, E. H. November 1978. "Socioeconomic and Institutional Restraints to Energy Development." *American Behavioral Scientist* 22:213–36.

Stobaugh, Robert, and Yergin, Daniel, eds. 1979. *Energy Future: Report of the Energy Project of the Harvard Business School*. New York: Random House.

Thumann, Albert, ed. 1981. *The Emerging Synthetic Fuel Industry*. Atlanta, Ga.: Fairmont Press.

Tierney, J. T. Spring 1984. "Government Corporations and Managing the Public's Business." *Political Science Quarterly* 99:73–92.

U.S. Atomic Energy Commission. 1974. *Coal Processing: Gasification, Liquefaction, Desulfurization: A Bibliography, 1930–1974*. Oak Ridge, Tenn.: U.S. Atomic Energy Commission, Office of Information Services, Technical Information Center.

U.S. Comptroller General. 1976. *Status and Obstacles to Commercialization of Coal Liquefaction and Gasification*. RED–76–81. Washington, D.C.: General Accounting Office.

———. August 17, 1977. *First Attempt to Demonstrate a Synthetic Fossil Energy Technology—A Failure*. EMD–77–59. Washington, D.C.: General Accounting Office.

———. 1982. *Evaluation of Administrative Procedures at the Synthetic Fuels Corporation*. GAO/RCED–83–27. Washington, D.C.: General Accounting Office.

Vietor, Richard H. K. Spring 1980. "The Synthetic Liquid Fuels Program: Energy Politics in the Truman Era." *Business History Review* 54:1–34.

———. 1984. *Energy Policy in America since 1945*. New York: Cambridge University Press.

Weitzman, M. L.; Newey, W.; and Rabin, M. Autumn 1981. "Sequential R&D Strategy for Synfuels." *Bell Journal of Economics* 12:574–90.

Welles, Chris. 1970. *The Elusive Bonanza*. New York: E. P. Dutton.

Whitehurst, Diane; Mitchell, Thomas O.; and Fancasiu, Malvina. 1980. *Coal Liquefaction: The Chemistry and Technology of Thermal Processes*. New York: Academic Press.

Xander, J. A., et al. July 1984. "Coal Gasification: An Economic Evaluation." *Technological Forecasting and Social Change* 25:309–27.

Index

Contributors

MICHAEL M. CROW is an assistant professor of political science and the assistant director of policy analysis at the Ames Laboratory at Iowa State University. His main teaching area is science policy. He has been involved in energy research for a number of years and is the author of several publications in the field of energy policy and research policy.

CYNTHIA M. DUNCAN is associate director of the Rural Economic Policy Program at the Aspen Institute. For three years she was research director at the Mountain Association for Community Economic Development in eastern Kentucky, where she directed research on economic development, policy toward the coal industry, and the rural working poor. She has written and published a variety of scholarly and popular papers on poverty and inequality, economic development, and the coal industry's effect on coalfield communities.

WILLIAM C. GREEN is an assistant professor of government at Morehead State University and a research associate with the Institute of Mining and Minerals Research at the University of Kentucky. He holds a Ph.D. in political science from the State University of New York at Buffalo and a law degree from the University of Kentucky College of Law. His teaching and research interests are in the areas of law and public policy, including energy, environment, and civil liberties policies. He is a frequent contributor to law journals on topics in these areas.

PATRICK W. HAMLETT is an associate professor of political science at the University of Missouri at Rolla. His chief research interests lie in science and technology issues. He has published several articles in diverse scholarly journals, including *Science, Technology and Human Values, Technology in Society*, and *Politics and the Life Sciences*.

F. LARRY LEISTRITZ is a professor of agricultural economics at North Dakota State University. He is the author of many articles and several books on environmental and socioeconomic impact assessments of energy and other large-scale development projects.

JOHN C. MITCHELL is Commissioner of Energy Research and Development within the Kentucky Energy Cabinet. He joined state government in 1977 and for several years served as manager of the Kentucky Demonstration Program throughout much of the synfuels era. He holds an M.S.M.E from the University of Kentucky and is a professional engineer specializing in coal-conversion technologies.

STEVE H. MURDOCK is a professor and head of rural sociology at Texas A & M University. His teaching and research interests lie in a number of areas, including social impact assessment and impact management in various policy areas.

HERBERT G. REID is a professor of political science at the University of Kentucky. He teaches and does research in contemporary political philosophy, American political thought and culture, Appalachian politics, and problems of ideology and change in advanced capitalist societies. His many articles have appeared in leading journals of political and social theory. He is a member of the editorial board of *Human Studies* and also serves on the executive committee of the Society for Phenomenology and the Human Sciences.

JOSEPH R. RUDOLPH is an associate professor of political science at Towson State University. His teaching and research interests lie predominantly in the area of comparative policy studies, especially in the field of energy. He has published in this area in a number of scholarly journals.

ANN R. TICKAMYER is an associate professor of sociology at the University of Kentucky. Her research interests are in the areas of stratification by gender, class, and region, and policy issues involving poverty and inequality. She is currently engaged in research on work and poverty in rural labor markets and is principal investigator of a study of gender differences in managerial careers funded by the National Science Foundation.

RICHARD H. K. VIETOR is a professor in the Department of Management at the Harvard Graduate School of Business Administration, where he teaches courses on international political economy and on the regulation of business.

After receiving a Ph.D. from the University of Pittsburgh in 1975, he taught at the University of Missouri before going to Harvard as the Newcomen Fellow in 1978. He is the author of numerous articles, cases, and books, including, most recently, *Energy Policy in America Since 1945* and *Telecommunications in Transition*.

THOMAS J. WILBANKS is the associate director and head of programs and planning in the energy division of Oak Ridge National Laboratory. He holds a Ph.D. in geography from the Maxwell School at Syracuse University. Formerly a faculty member at Syracuse University and the University of Oklahoma, he has been involved at Oak Ridge in technology and policy assessments of synthetic fuels since 1974.

SABRINA WILLIS is a graduate student in the Department of Government at the University of Virginia. She has written or coauthored several papers dealing with synthetic fuels development in national and international perspective.

ERNEST J. YANARELLA is a professor of political science at the University of Kentucky. He teaches political theory and public policy (including energy and the environment, arms control and defense policy, and agricultural policy). His work in critical policy studies is reflected in his two dozen scholarly articles and three previous books. He is currently completing a book on contemporary science fiction and the ecological imagination.